D0882685

Machinists' Ready Reference Manual

John E. Traister

McGraw-Hill Book Company
New York   St. Louis   San Francisco   Auckland
Bogotá   Hamburg   London   Madrid   Mexico
Milan   Montreal   New Delhi   Panama
Paris   São Paulo   Singapore
Sydney   Tokyo   Toronto

Library of Congress Cataloging-in-Publication Data

Traister, John E.
  Machinists' ready reference manual.

  Includes index.
  1. Machine-shop practice—Handbooks, manuals, etc.
I. Title.
TJ1165.T65 1987   621.9′02   87-2801
ISBN 0-07-065143-4

  234567890   KGP/KGP   8921098

ISBN 0-07-065143-4

*The editors for this book were Harold B. Crawford and Nancy Young,
the designer was Naomi Auerbach, and the production
supervisor was Richard A. Ausburn. It was set in Century Schoolbook
by University Graphics, Inc.*

Printed and bound by The Kingsport Press.

Contents

Preface   xi

CHAPTER 1. Math for the Machinist   1

Types of Calculators   2
Solving Machine-Shop Problems with the Electronic
  Calculator   5
Taper Angle   9
More Trigonometric Functions   17
Computer Use   18

CHAPTER 2. Mathematical Data and Equations   25

Equations   25
Meanings of Symbols   27

CHAPTER 3. Conversion Tables   37

CHAPTER 4. The Metric System   **43**

Rules for Writing Metric Quantities   **44**
Pronunciation of Metric Terms   **54**
Weight, Mass, and Force   **55**
Typewriting Recommendations   **56**
Longhand   **58**
Shorthand   **58**

CHAPTER 5. Machine-Shop Drawings   **59**

Types of Drawings   **60**
Schedules   **73**
Symbols   **73**
Drawing Dimensions   **75**
Applications   **75**
Working Drawings   **82**

CHAPTER 6. Measuring Tools and Measurements   **91**

Steel Rules   **93**
Hermaphrodite Caliper   **95**
Micrometers   **97**
Angles and Angular Measurements   **104**
Vernier Height Gage   **108**
Dial Indicators   **111**
Angle Gage Blocks   **113**
Measuring Threads   **115**
Gage Blocks   **119**

CHAPTER 7. Machine-Shop Materials   **121**

Carbon Steels   **124**
Alloy Steels   **125**
High-Strength Low-Alloy Steels   **125**
Stainless Steels   **125**
Tool-and-Die Steels   **126**

CHAPTER 8. Soldering, Brazing, and Welding   **131**

Soldering   **131**
Brazing (Hard Soldering)   **133**
Welding   **134**

CHAPTER 9. Heat Treating Steel   **173**

Hardening   **173**
Tempering   **174**
Annealing   **175**
Forging   **175**
Normalizing   **175**
Carburizing   **176**
Heat for Heat Treating   **177**
Thermocouple Pyrometers   **177**
Hardness Tester   **187**

CHAPTER 10. Metal Finishes   **181**

Damascening Metal   **181**
Blueing   **183**
Parkerizing   **188**

Plating Metal   189
Metal Polishing   192

**CHAPTER 11.   Metalworking Coolants and
Lubricants**   195

Selecting Cutting Fluids   196
Application of Cutting Fluids   202
Cutting Fluid Considerations   202
Applications of Cutting Fluids   203
Solid Lubricants   205

CHAPTER 12.   Bench Work and Tools   207

Layout Work   208
Bench Tools   210
Hand Broaching   219
Hand Scrapers   219

CHAPTER 13.   Principles of Lathe Operation 221

Parts of a Lathe   222
Installation   224
Centering Work   227
Securing Workpiece in Lathe   232
Lathe Cutting Tools   234
Special Lathe Work and Tools   234
Taper Turning   240

CHAPTER 14. Cutting Tools   251

Lathe Cutting Tools   251
Grinding Lathe Tools   253
Milling Cutters   258
Reamers   259

CHAPTER 15. Taps and Dies   261

Taps   261
Dies   263
Metric Threads   269
Tap Extractors   269

CHAPTER 16. Drilling Operations   271

Twist Drills   271
Radial Drilling   274
Drilling in the Lathe   281

CHAPTER 17. Shapers and Planers   283

Shaper Operation   285
Control of Machine Tools   287

CHAPTER 18. Milling   289

Milling Application   292
Milling in the Lathe   296

CHAPTER 19. Dividing Heads and Indexing   299

Compound Indexing   300
Differential Indexing   300
Angular Indexing   301
Block Indexing   301
The Sector Arms   306
Mounting the Attachment   308
Adjustments   311

CHAPTER 20. Arbors, Collets, and Adapters   313

Collets   317
Lathe Adapters   318

CHAPTER 21. Grinders and Grinding   321

Abrasives   321
Precision Grinding   329
Grinding in the Lathe   331
External Grinding   333
Internal Grinding   334
Grinding Valves   335
Grinding Flat Valves   335
Grinding 60 Degree Lathe Centers   336
Surface Grinding   336
Selected Charts and Tables Used in Grinding
Applications   340

CHAPTER 22. Laps and Lapping   341

Lapping Compounds   341
Lapping Operations   346

Practical Applications   348
Lapping Rifle Barrels   348

CHAPTER 23. Punches and Shears   355

Punches   355
Shears and Snips   357

CHAPTER 24. Broaching   361

CHAPTER 25. Metal Spinning   367

Spinning Operation   367
Spinning Tools   368

CHAPTER 26. Gears   373

Gear Nomenclature   373

CHAPTER 27. Threads and Threading Operations   379

Thread Forms   381
Cutting Threads on a Lathe   390

CHAPTER 28. Hand Files and Filing   393

File Features   393
Types of Files   396

File Selection   398
Using Files   399
The Care of Files   404

CHAPTER 29. Mechanics   407

Mechanical Power   407
Electric Power   408
Electric Motors   409
Machine Drives   412
Selecting the Motor for the Job   415

Glossary 425
Index   461

Preface

This is a concise, practical, and useful manual designed for easy reference by any machinist, sheet-metal worker, machine designer, or metalworking specialist. Users of this manual will save time and money by applying its information to everyday jobs.

The topics covered in the manual include math for the machinist; using electronic calculators to solve shop problems; computer usage in metalworking; equations; conversion tables; the metric (SI) system; machine-shop drawings; measuring tools; machine-shop measurements; materials for metalworking—steels of various types: carbon, alloy, high-strength, stainless, tool-and-die—soldering, brazing, and welding; heat treating steel; metal finishes; metalworking coolants and lubricants; bench work and tools; principles of lathe operation; cutting tools; taps and dies; drilling operations; shapers and planers; milling; dividing heads and indexing; arbors, colletts, and adapters; grinders and grinding; laps and lapping; punches and shears; broaching; metal spinning;

gears; threads and threading operations; hand files and filing; and mechanics.

Each section covers its topic in great practical detail. The user of the manual is given specific hands-on tips which help the reader perform a given task better and more efficiently. There is no other work that gives such carefully detailed directions and hints for machine-shop workers.

A large number of tables and illustrations help the reader do more in less time. These tables and illustrations give actual usable data for direct application to specific machine-shop jobs. Thus, the user can keep this manual alongside a metalworking machine or on the bench for constant reference. Such a reference work has long been needed in all types of machine shops—mass-production, "one-off," and job shops.

Every machine-shop worker—from the newest beginner to the most experienced journeyman—will find the contents of this manual helpful. Since every major machine is covered, machine operators will gain from use of the manual. Thus, lathe operators, milling-machine operators, grinder operators, and broaching-machine operators—among others—will get tips on how to get more from their machines.

Practical data from machine manufacturers is included in many of the tables and illustrations. Since the builders of metalworking machines know the work capabilities of their units better than anyone else, this information is most valuable to operators.

The author hopes that every user of this manual finds ideas, tips, or methods that will help on the job. To get the most from this manual, make a habit of referring to it whenever a question arises about a method or machine. You will find the book is a constant source of useful information.

John E. Traister

1

Math for the Machinist

Machinists of the past have rated the slide rule as one of their most important instruments for making the many necessary math calculations required in the shop. The slide rule did help reduce long minutes of paper-and-pencil calculations to a few simple manipulations of the "slipstick" and "runner." While the slide rule has proved indispensable since its invention in the 1850s, the electronic pocket calculator is faster, more versatile, more compact, more accurate, and better able to solve today's machineshop problems involving mathematical calculations.

There is no need to explain the fundamentals of electronic calculator operation since this information may readily be obtained from the handbook accompanying the devices and since practically anyone can master the basic operations of an electronic calculator in a single evening. The paragraphs that follow will explain how to make specific basic machine-shop calculations. Actual directions are given for pressing the required keys in most cases.

A selected number of examples have been chosen. These are basic; by no means do they attempt to cover all of the possible uses of the calculator. Other examples solvable by the same processes should readily occur to the reader.

TYPES OF CALCULATORS

The slide rule calculator.   The "slide rule calculators" (Fig. 1.1) were developed to replace the conventional slide rule. These compact tools perform calculations of roots, powers, reciprocals, common and natural logarithms, and trigonometry in addition to basic arithmetic. Versatile memory functions may include Store, Recall, Sum to Memory, and Memory/Display Exchange. Besides these, answer display to 10 places to the right of the decimal sign in standard format or in scientific notation, correlation, linear regression, trend-line analysis, and more are available in these units.

Programmable calculators.   Programmable calculators (Fig. 1.2) are really miniature computers—with powerful capabilities. Multiuse memories provide addressable memory locations in which data can be stored for recall. Several hundred program steps or up to 100 memories are available.

Master library modules.   The master library module available from Texas Instruments, Inc., plugs into the electronic calculator and includes many different programs in key areas.

Solid-state software libraries.   Solid-state software is an advance in micromemory technology and provides new programming versatility. It only takes seconds and a few key strokes to drop in a module and access a program. Optional solid-state software library modules are currently not available for strictly machine-shop use.

FIG. 1.1 The Texas Instruments Model TI-55 electronic slide-rule calculator. *(Texas Instruments.)*

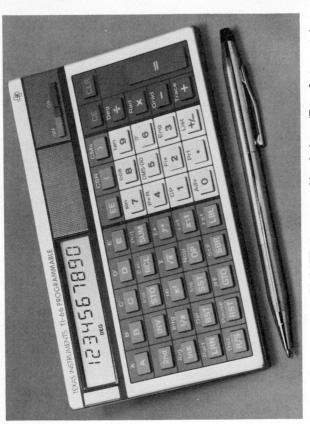

FIG. 1.2 Texas Instruments Model TI-66 programmable calculator. *(Texas Instruments.)*

Printing calculators. Several calculators are available with printing attachments which imprint all actions of the calculator on a tape. They are useful when the machinist needs to refer to notes while at a machine or the bench.

The type of electronic calculator most often found in the machine shop is the hand-held type, costing less than $100. The examples to follow will be limited to that type. However, any calculator capable of the four basic mathematical functions plus the other functions previously described for the slide rule calculator may be employed. Any deviations caused by the difference in models should easily be corrected by referring to the owner's manual.

SOLVING MACHINE-SHOP PROBLEMS WITH THE ELECTRONIC CALCULATOR

The following equation may be used to calculate cutting speeds of various machine shops tools such as lathes and milling machines:

Cutting speed = 3.1416 × revolutions per minute
  × diameter of work / 12

For example, to find the cutting speed for a piece of work which is ½ in in diameter and turning at 480 revolutions per minute (rpm):

If a certain cutting speed is desired, the required rpm of the machine may be found by the equation

rpm = 12 × cutting speed / 3.1416 × diameter of work

Enter	See displayed	
1. Key in pi or 3.1416	3.1416	
2. Press the multiplication key, ×	3.1416	
3. Key in rpm	480	
4. Press multiplication key, ×	1507.9680	
5. Key in diameter of work, .5 in	0.5	
6. Press the division key, ÷	753.9840	
7. Key in 12	12	
8. Press equals key, =	62.8320	[feet per minute (ft/min)]

For example a cutting speed of 40 ft/min is desired on a piece of work that has a diameter of 2 in. To find the machine's speed:

Enter	See displayed	
1. Enter the number 12	12	
2. Press the multiplication key, ×	12.0000	
3. Enter cutting speed	40	
4. Press the division key, ÷	480.0000	
5. Enter pi or 3.1416	3.1416	
6. Press the division key, ÷	152.7883	
7. Enter diamter of work	2	
8. Press equals key, =	76.3941	(rpm)

To determine the time in minutes required to take one complete cut on a workpiece, the following equation may be used:

Time in minutes = total length of cut in inches / rpm
× feed per revolution

For example, to find the time required to make a cut 24 in long if the speed of the machine is 80 rpm and the feed is 0.010 in:

Enter		See displayed
1. Enter length of piece	24	
2. Press division key, ÷	24.0000	
3. Enter rpm	80	
4. Press division key, ÷	0.3000	
5. Enter rate of feed	0.010	
6. Press equals key, =	30.000	(minutes required to perform the work)

Tapers often need to be turned in the lathe, and each type must be calculated precisely to obtain the desired dimensions. Depending on many conditions, one of the following equations may be used:

Taper per inch = taper per foot / 12
Taper per foot = 12(lg. dia. − sm. dia.) / length of taper
Taper in any length = length of taper × taper per foot / 12
Length of taper = 12(lg. dia. − sm. dia.) / taper per foot

When turning a taper in a metal-cutting lathe, either the compound rest or a special taper-turning attachment may be used. However, the machinist will usually set the tail stock over the desired distance. For example, a shaft, 24 in in length, is to have a taper at one end of 1.33 in and 0.70 in at the other. To find the taper per foot:

The tail stock of a lathe can be set over or offset by means of

Enter		See displayed
1. Key in large diameter	1.33	
2. Press equals key, =	1.3300	
3. Key in smaller diameter	0.7	
4. Press minus key, —	0.6300	
5. Press multiplication key, ×	0.6300	
6. Enter 12	12	
7. Press equals key, =	7.5600	
8. Press division key, ÷	7.5600	
9. Enter length of shaft	24	
10. Press equals key, =	0.3150	(inches of taper per foot)

screws in the tail stock. The equation for determining the amount of offset is

Offset = lg. dia. — sm. dia. / 2

For example, to find the amount of tail stock offset required to taper a 12-in bar with 1.5 in at one end and .75 in at the other:

Enter		See displayed
1. Enter the large diameter	1.5	
2. Press equals key, =	1.5000	
3. Enter the small diameter	0.75	
4. Press the minus key, —	0.75	
5. Press the division key, ÷	0.75	
6. Enter 2 (in equation)	2	
7. Press equals key, =	0.3750	(inches)

Therefore, the tail stock should be offset 0.3750 in to obtain the desired taper.

The above equation can be modified slightly to calculate the tail stock offset when the offset runs for only part of the total length of the shaft:

Offset = (lg. dis. − sm. dia.) × total length / 2 × length of taper

For example, a rifle barrel, 20 in in length, is to be tapered starting 3 in from the chamber end, which is .938 in in diameter, and ending at a diameter of .700 in at the muzzle; the total length of offset is therefore 17 in. To find the required offset of the tail stock:

Enter	See displayed	
1. Enter large diameter	0.937	
2. Press equals key, =	0.9370	
3. Enter small diameter	0.700	
4. Press multiplication key, ×	0.7000	
5. Enter total length of barrel	20	
6. Press division key, ÷	14.0000	
7. Enter 2 (in equation)	2	
8. Press division key, ÷	7.0000	
9. Enter length of offset	17	
10. Press equals key, =	0.4117	(inches)

The offset required at the tailstock to achieve the desired taper is 0.4117 in.

TAPER ANGLE

A very steep taper (usually over 10 degrees with the centerline) is usually called an angle rather than a taper. Figure 1.3 shows a lathe center with the various angles. Referring to this sketch, the

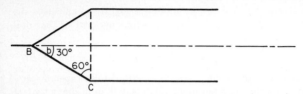

FIG. 1.3 Typical lathe center, showing the various angles.

following equation may be used to determine the angle *b* with the centerline:

$$\tan b = \text{taper per foot} / 24$$

When the diameters and length of the taper are known, the following equation may be used:

$$\tan b = \text{lg. dia.} - \text{sm. dia.} / 2 \text{ length of taper}$$

For example, if the taper per foot is .55 in, to find the angle with the centerline and the included angle:

Enter	See displayed
1. Enter the taper per foot	0.55
2. Press equals key, =	0.5500
3. Press division key, ÷	0.5500
4. Enter 24 (in equation)	24
5. Press equals key, =	0.0229

From a table of tangents, .0229 = 1 degree, 41 minutes, which makes the included angle:

$$2 \times b \quad \text{or} \quad 3 \text{ degrees, 22 minutes}$$

A short taper may be made with the lathe compound rest (Fig. 1.4). The tail stock setover is the most popular way of tapering in the lathe (Fig. 1.5).

Screw threads are familiar to every machinist, and the two basic types are outside threads (as may be seen on a machine screw or bolt) and inside threads (as may be seen on a nut, pipe coupling, etc.). Detailed information is given on screw threads in Chap. 27 in this book. However, the following will serve to demonstrate a few uses of the pocket calculator in making threading calculations.

The depth of a v thread is found by the equation

Depth = .866 / number of threads per inch

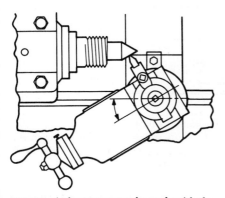

FIG. 1.4 A short taper may be made with the lathe compound rest.

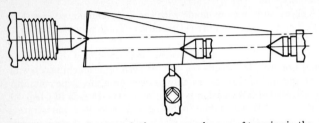

FIG. 1.5 Tail-stock setover is the most popular way of tapering in the lathe.

For example, to find the depth of a v thread with 20 threads per inch:

Enter	See Displayed
1. Enter 0.86682	0.866
2. Press division key, ÷	0.8660
3. Enter threads per inch	20
4. Press equals key, =	0.0433

Therefore, the threads should be 0.0433 in deep.

The angle of helix is the angle which a spiral makes with the axis of the work. When using, say, a milling machine, the table must be set at this angle when cutting a helix. This angle may be found by the equation

Tangent of helix angle = 3.1416 × diameter of work / lead of helix

For example, a helix with a 20-in lead is to be cut on a piece of work 2 in in diameter. To find the angle of helix:

Enter	See displayed
1. Enter 3.1416	3.1416
2. Press multiplication key, ×	3.1416
3. Enter diameter of work	2
4. Press division key, ÷	6.2832
5. Enter lead	20
6. Press equals key, =	0.3141

Using trigonometric functions, it may be found that tangent 0.3141 = 17 degrees, 26 minutes—or just a fraction less. The scale on the milling machine would probably be set at 17½ degrees unless ultraprecision became necessary.

The angle of helix may be found graphically on paper and then cut out and wrapped around the work with the side representing the circumference encircling the work. The hypotenuse of the triangle will trace out the helix on the work. The triangle chart in Fig. 1.6 will be of assistance in finding the angle of helix as well as other angular solutions, including the rapid solution of right-angle and oblique-angle triangles.

There are numerous types and sizes of gears, and most are familiar to all machinists. The four general types are spur gears, bevel gears, worm gears, and spiral or helical gears.

One problem encountered when working with gears is the number of teeth in a particular gear. One equation is

Number of teeth = diametral pitch × pitch diameter

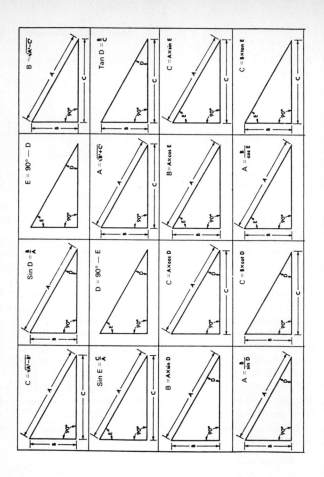

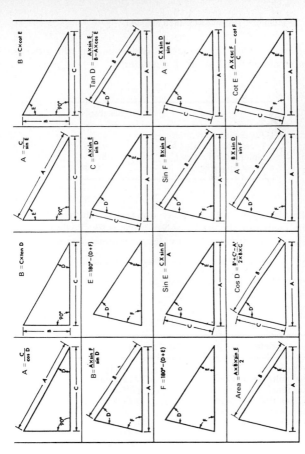

FIG. 1.6 Triangle chart for the rapid solution of right-angle and oblique-angle triangles.

For example, to find the number of teeth in a gear with a diametral pitch of 4 and an 8-in pitch diameter:

Enter	See displayed
1. Enter diametral pitch	4
2. Press multiplication key, ×	4.0000
3. Enter pitch diameter	8
4. Press equals key, =	32.0000

Therefore, a gear with the given dimensions will have 32 teeth.

Grinding wheels are often used to finish work. To find the surface speed of a grinding wheel in feet per minute, use the equation

Surface speed = 3.1416 × rpm × diameter in inches / 12

For example, to find the surface speed of a 10-in grinding wheel revolving at a speed of 1750 rpm:

Enter	See displayed	
1. Enter pi	3.1416	
2. Press multiplication key, ×	3.1416	
3. Enter rpm	1750	
4. Press multiplication key, ×	5497.8000	
5. Enter diameter of wheel	10	
6. Press division key, ÷	54978.0000	
7. Enter 12 (in equation)	12	
8. Press equals key, =	4581.5000	(feet per minute)

MORE TRIGONOMETRIC FUNCTIONS

The three trigonometric functions, sine, cosine, and tangent, are used frequently in making all sorts of machine shop calculations. The relationships are related to the right triangle as shown in Fig. 1.7 and are defined as follows

Sin = length of opposite side / length of hypotenuse = O / H
Cos = length of adjacent side / length of hypotenuse = A / H
Tan = length of opposite side / length of adjacent side = O / A

The Sin Cos and Tan keys on a pocket calculator each assume there is an angle in the display in units specified by the setting of the Drg key. When any of these three keys is pressed, the appropriate function (sine, cosine, or tangent) of the displayed angle is computed and displayed immediately. These three keys, on most calculators, do not affect calculations in progress and can be used at any time.

To illustrate the use of these keys, compute the sine, cosine, and tangent of 90 degrees and 90 grads:

Press	Display (comments)
Off, On/C	(This makes certain the calculator is in degree mode.)
90, Sin	1.
90, Cos	0.
90, Tan	ERROR (The tangent of 90° is undefined.)
On/C	(Clears error condition.)
Drg, Drg	"0. (Converts to grads mode.)
90, Sin	".98768834
90, Cos	".15643447
90, Tan	"6.3137515

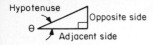

FIG. 1.7 Relationships of the sine, cosine, and tangent in the right triangle.

The examples shown were performed on an Unitrex Instant calculator, which is about the simplest one made. Slight changes may be necessary for other models, but the instructions included with them all should readily show what changes (if any) are necessary.

Practical applications requiring mathematical calculations are found in the chapters to follow—all of which can be adapted and worked on the pocket calculator or "electronic slide rule." While the explanations will be in a different format (directions for pressing each key will not be given), anyone familiar with the data presented in this chapter should find little difficulty in applying the equations to his or her own calculator—especially if the owner's manual is referred to when a review of the calculator functions becomes necessary.

COMPUTER USE

The use of computers has now reached the machine-shop floor. To use a computer, it is *not* necessary to understand exactly how it operates. But the user must know how to provide instructions in a language the computer understands.

A set of instructions for a computer is called a "program." Each program consists of a series of steps which are performed in a certain order. Then the computer will follow the instructions and act upon what it has been instructed to do. A typical computer suitable for machinist's calculations is shown in Fig. 1.8.

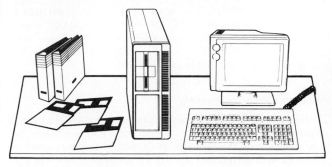

FIG. 1.8 A Wang Professional Computer is used extensively by the author for numerous applications, including machinist's calculations.

BASIC is one language that most computers understand. It stands for Beginner's All-Purpose Symbolic Instruction Code, and there is hardly a computer, big or small, which cannot communicate in BASIC.

To make BASIC work, the computer must also have another program or instruction in operation that can interpret the words it reads, and then perform the operations which are desired by the programmer; it's much like having an interpreter in a foreign country. But the computer is not quite as smart as your average human interpreter. It's fast but by human standards quite stupid. For example, if you tell the computer to GO TO, it probably will not do anything except print ERROR or something to that effect—depending upon which computer you are using. Had you typed GOTO, the computer would immediately know what to do. In other words, in BASIC, one space out of place can make or change a program completely.

Even though BASIC has been designed to "understand" English, it has a limited vocabulary. For example, BASIC knows what to do when STOP is typed on the keyboard, but type WHOA, WAIT A MINUTE, HALT, or other synonym for STOP and the computer does not know what to do.

Preliminary Program Planning

To write any computer program, a precise definition of what the program is to accomplish is required. Also, a detailed, step-by-step plan of how the program goals will be achieved must be known or worked out, and finally, a knowledge of how the programming statements can be used to complement the step-by-step plan is necessary. The following steps should be taken:

1. The precise definition of what the program is to accomplish is the first and most important part of the creative cycle. This step must describe exactly what is to be accomplished.
2. Work out a plan (program), in a language that the computer understands, to accomplish the goal step in no. 1 above.
3. Run the program to test for accuracy.
4. Debug program so that it operates properly on your computer.
5. Store or save program for future use so that it can be recalled at any convenient time.

One problem that often confronts machinists is to calculate the amount of tail stock setover required for taper turning. When the taper runs the entire length of the workpiece, most machinists subtract the small diameter from the large diameter and then divide by 2 to determine the amount of tail stock offset. This calculation can be performed quickly on a $10 pocket calculator, but since this is a familiar calculation, it will be used to demonstrate how a computer program is developed.

The exact method of developing a step-by-step computer program will vary from person to person. Some people rely heavily on a technique known as flow charting, while others have an intuitive "feel" for program development and can write programs right off the tops of their heads.

Writing the Program

The first step is to begin by using a REM, or remark statement, to identify the program:

```
10 REM **AMOUNT OF OFFSET FOR TAPER TURNING**
```

The next step involves inputting the values of the items to be used in the computation. This data will come from the keyboard, and a name is needed for the variable that will receive the data. Here, A is used for the large diameter, and B is used for the small diameter. At this point, the program will look something like this:

```
10 REM **AMOUNT OF OFFSET REQUIRED FOR TAPER TURNING**
20 INPUT A
30 INPUT B
```

Each program line is numbered; this is necessary to give the computer directions. The input is needed from the user, so the computer will ask a question on the screen by displaying a question mark. Since the amount of tail stock setover is to be found, the following equation must be entered in the program:

$$O = D - d / 2$$

However, since most computers read upper- and lowercase letters the same, a change has been made to suit the computer. The letter D in the equation represents the large diameter of the workpiece, while d represents the smaller diameter. In the program, we have used A and B respectively to represent these diameters.

In the next line (40), the equation is written as O = A − B / 2, and most of the program is now complete. However, to make the program useful, we want to see the results, so the computer is told to display the results on the screen via the instruction PRINT. In fact, it is desired to see all the figures: both diameters and the amount of tail stock setover. The instructions are given as follows:

```
50 PRINT A
60 PRINT B
70 PRINT O
80 STOP
```

The entire program will then appear a shown in Fig. 1.9, and when this program has been keyed into the computer, all that is necessary is to type RUN and, assuming that a taper is to have a large diameter of 1.5 in and a small diameter of 0.87 in, the screen

```
10 REM **AMOUNT OF OFFSET REQUIRED FOR TAPER TURNING**
20 INPUT A
30 INPUT B
40 O=A-B/2
50 PRINT A
60 PRINT B
70 PRINT O
80 STOP
```

FIG. 1.9 Simple BASIC program used to calculate tail-stock setover.

```
Ok
RUN
? 1.5
? 0.87
 1.5
 .87
 1.065

RUN "PRNSCRN
```

FIG 1.10 View of computer 'screen' when the program in Fig. 1.9 is run.

---

will look like Fig. 1.10 when the program is run. When the first ? appears, enter the large diameter, 1.5, and press Return; when the second ? appears, enter the small diameter, .97, and press Return. The computer will make the calculation and display the answer on the screen as 1.065.

Now this program is written in the simplest form possible and is perfectly all right for one-time use. However, if it becomes necessary to refer to the program at some future date, there is a good chance of forgetting what the computer is asking for first. Was it the large or small diamter? The original program, however, can

```
10 REM **AMOUNT OF OFFSET FOR TAPER TURNING**
20 REM **WHEN TAPER RUNS ENTIRE LENGTH OF BAR**
30 INPUT "DIAMETER IN INCHES AT LARGE END OF TAPER";A
40 INPUT "DIAMETER IN INCHES AT SMALL END OF TAPER";B
50 O=A-B/2
60 PRINT "A TAPERED BAR WITH A DIAMETER OF";A;"INCHES AT LARGE END"
70 PRINT "AND A DIAMETER OF";B;"INCHES AT SMALL END"
80 PRINT "REQUIRES A TAILSTOCK OFFSET OF";O;"INCHES"
```

FIG. 1.11 A more sophisticated version of the program in Fig. 1.9.

```
Ok
RUN
DIAMETER IN INCHES AT LARGE END OF TAPER? 1.5
DIAMETER IN INCHES AT SMALL END OF TAPER? 0.87
A TAPERED BAR WITH A DIAMETER OF 1.5 INCHES AT LARGE END
AND A DIAMETER OF .87 INCHES AT SMALL END
REQUIRES A TAILSTOCK OFFSET OF 1.065 INCHES
Ok
RUN "PRNSCRN
```

FIG. 1.12 View of computer 'screen' when the program in Fig. 1.11 is run.

easily be dressed up to eliminate any of these problems. The program shown in Fig 1.11 will handle the job and the printout in Fig. 1.12 is what you will see once the computer has finished its calculation.

2

Mathematical Data and Equations

EQUATIONS

An equation is a brief statement of a rule, in which letters or other symbols are used to denote the different quantities involved. For example, the rule for finding the volume of a rectangular prism is as follows:

> The volume of a rectangular prism is equal to the product of the length, width, and height of the prism. If the dimensions of the prism are taken in inches, the volume will be in cubic inches; if they are taken in feet, the volume will be in cubic feet; and so on. However, if the volume is denoted by V, the length by l, the width by W, and the height by h, then the rule may be stated by the equation,
>
> $$V = l \times W \times h$$
>
> This equation indicates that the volume V is equal to the product of the length l, the width W, and the height h of the prism. Where

several letters are multiplied in an equation, it is customary to omit the multiplication signs—the multiplication then being taken for granted. The foregoing equation, therefore, would ordinarily be written $V = lWh$.

Note from the preceding that an equation consists of two parts separated by the sign of equality. The letters or symbols denoting the quantities that are known are usually placed at the right of the equality sign, and the letter or symbol designating the value to be found is placed at the left of the equality sign. To apply an equation to the solution of a problem, a numerical value is substituted for each letter that denotes a known quantity, and the indicated mathematical operations are then performed. Care must be observed, however, in using an equation, to have all weights, dimensions, or other values expressed in the units required by the equation.

Letters with additional marks, such as C', a'', d_1, T_a, etc., are often found in equations when similar quantities are to be represented by the same letter and yet to be distinguished from one another. The marks, $'$ and $''$ are termed prime and second, respectively, and the marks $_1$ and $_a$ are termed subscripts or "subs." The four examples just given are read "large C prime", "a second", "d sub one", and "large T sub a."

Parentheses ( ) and brackets [ ] are used in equations to indicate that the quantities enclosed by them are to be subjected to the same operation. The sign $-$ before an expression in parentheses or brackets affects the entire expression, and, if the parentheses or brackets are removed, the signs $+$ and $-$ within them must be interchanged. However, if the sign $+$ precedes the brackets, they may be removed without changing any signs.

The expression, $212 - (36 + 75 - 49)$, for example, becomes

$212 - 36 - 75 + 49$ when the parentheses are removed, but the removal of the brackets from, say, the expression $65 + [20 + 9 - 14]$ gives $65 + 20 + 9 - 14$.

In most cases, the multiplication sign is omitted before and after parentheses or brackets, and also before the radical sign used to express the square root of a number or group of numbers.

The following example shows how the basic operations are performed. Find the volume of a block of machined steel 28 in long, 15 in wide, and 12 in high. The solution is to apply the equation $V = lWh$ and substitute the given values in the equation:

$$V = 28 \times 15 \times 12 = 5040 \text{ in}^3$$

Basic mathematical operations should not need reviewing, as they are used daily in machine-shop work. However, certain functions such as determining the square root often need reviewing. Reference material should be available for the mechanic. With the modern electronic pocket calculator, such functions as square roots, reciprocals, and the like are often performed as explained in Chap. 1.

MEANINGS OF SYMBOLS

In the following equations, the letters have the meanings here given, unless otherwise stated:

D = larger diameter
d = smaller diameter
R = radius corresponding to D
r = radius corresponding to d
p = perimeter or circumference

C = area of convex surface = area of flat surface that can be rolled into the shape shown

S = area of entire surface = C + area of the end or ends

A = area of plane figure

π = ratio of circumference of any circle to its diameter
= 3.1416   approximately

V = volume of solid

The other letters used will be found on the illustrations.

Triangle

Using letters to denote the angles,

$$D = B + C E + B + C = 180°$$
$$B = D - C E' + B + C = 180°$$
$$E' = E \qquad B' = B$$

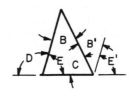

For a right triangle, c being the hypotenuse,

$$c = \sqrt{a^2 + b^2}$$
$$a = \sqrt{c^2 - b^2}$$
$$b = \sqrt{c^2 - a^2}$$

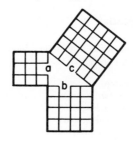

If c = length of side opposite an acute angle of an oblique triangle, and the distance e is known,

$$c = \sqrt{a^2 + b^2 - 2be}$$
$$h = \sqrt{a^2 - e^2}$$

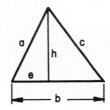

If c = length of side opposite an obtuse angle of an oblique triangle,

$$c = \sqrt{a^2 + b^2 + 2be}$$
$$h = \sqrt{a^2 - e^2}$$

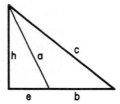

Any triangle inscribed in a semicircle is a right triangle, and

$$c{:}b = a{:}h$$
$$h = \frac{ab}{c} = \frac{ce}{a}$$
$$a{:}b + e = e{:}a = h{:}c$$

For any triangle,

$$A = \frac{bh}{2} = \frac{1}{2}bh$$

$$A = \frac{b}{2} \sqrt{a^2 - \left(\frac{a^2 + b^2 - c^2}{2b}\right)^2}$$

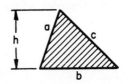

Also,

$$A = s(s - a)(s - b)(s - c)$$

in which

$$s = \frac{1}{2}(a + b + c).$$

Rectangle and Parallelogram

$$A = ab$$
$$A = b\sqrt{c^2 - b^2}$$

Trapezoid

$$a = \frac{1}{2}h(a + b)$$

Trapezium

Divide the figure into two triangles and a trapezoid; then,

$$a = \tfrac{1}{2}bh' + \tfrac{1}{2}a(h' + h) + \tfrac{1}{2}ch$$
$$A = \tfrac{1}{2}\,[bh' + ch + a\,(h' + h)]$$

Or divide into two triangles by drawing a diagonal. Considering the diagonal as the base of both triangles, call its length l, and call the altitudes of the triangles h_1 and h_2; then,

$$A = \tfrac{1}{2}l\,(h_1 + h_2)$$

Regular Polygon

Divide the polygon into equal triangles and find the sum of the partial areas. Otherwise, square the length of one side and multiply by proper number from the following table.

Name	No. of sides	Multiplier	Name	No. of sides	Multiplier
Triangle	3	.433	Heptagon	7	3.634
Square	4	1.000	Octagon	8	4.828
Pentagon	5	1.720	Nonagon	9	6.182
Hexagon	6	2.598	Decagon	10	7.694

Irregular Areas

Divide the area into trapezoids, triangles, parts of circle, etc. and find the sum of the partial areas.

If the figure is very irregular, the approximate area may be found as follows: Divide the figure into trapezoids by equidistant parallel lines b, c, d, etc., and measure the lengths of these lines. Then calling a the first and n the last length, and y the width of strips,

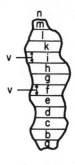

$$A = yr\left(\frac{a + n}{2} + b + c + \text{etc.} + m\right)$$

Sector

If l denotes the length of the arc, and E the angle in degrees* and decimals of a degree,

$$l = \frac{Er}{57.296} = .0175\ Er \quad \text{nearly}$$

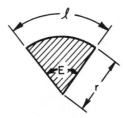

Then

$$A = \tfrac{1}{2}lr$$

$$A = \frac{\pi r^2 E}{360} = .008727 r^2 E$$

* If the angle E is stated in degrees, minutes, and seconds, the minutes and seconds must be reduced to decimals of a degree. To do this, divide the number of minutes by 60 and the number of seconds by 3600 and add the sum of the quotients to the number of degrees. Thus, $28°\ 42'\ 18'' = 28 + 42/60 + 18/3600 = 28 + .7 + .005 = 28.705°$.

Circle

$$p = \pi d = 3.1416d$$
$$p = 2\pi r = 6.2832r$$
$$p = 2\sqrt{\pi A} = 3.5449 \sqrt{A}$$
$$p = \frac{2A}{r} = \frac{4A}{d}$$
$$d = \frac{p}{\pi} = \frac{p}{3.1416} = .3183p$$
$$d = 2 \sqrt{\frac{A}{\pi}} = 1.1284 \sqrt{A}$$

$$r = \frac{p}{2\pi} = \frac{p}{6.2832} = .1592p$$
$$r = \sqrt{\frac{A}{\pi}} = .5642 \sqrt{A}$$
$$A = \frac{\pi d^2}{4} = .7854d^2$$
$$= \pi r^2 = 3.1416r^2$$
$$= \frac{pr}{2} = \frac{pd}{4}$$

Ring

$$A = \frac{\pi}{4r} (D^2 - d^2)$$

Segment

$$A = \tfrac{1}{2} [lr - c(r - h)]$$
$$= \frac{\pi r^2 E}{360} - \frac{c}{2} (r - h)$$
$$l = \frac{\pi r E}{180} = .0175rE$$
$$E = \frac{180l}{\pi r} = 57.2956 \frac{l}{r}$$

Ellipse

$$p^* = \pi \sqrt{\frac{D^2 + d^2}{2} - \frac{(D - d)^2}{8.8}}$$

$$A = \frac{\pi}{4} Dd = .7854 Dd$$

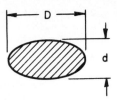

Cylinder

$$C = \pi dh$$
$$S = 2\pi rh + 2\pi r^2$$
$$= rdh + \frac{\pi}{2} d^2$$
$$V = \pi r^2 h = \frac{\pi}{4} d^2 h$$
$$V = \frac{p^2 h}{4\pi} = .0796 p^2 h$$

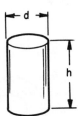

Frustum of Cylinder

h = ½ sum of greatest and least heights
$$C = ph = \pi dh$$
$$S = \pi dh + \frac{\pi}{4} d^2 + \text{area of elliptic top}$$
$$V = Ah = \frac{\pi}{4} d^2 h$$

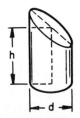

* The perimeter of an ellipse cannot be exactly determined without a very elaborate calculation, and this equation is merely an approximation giving close results.

Prism or parallelepiped

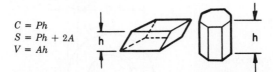

$$C = Ph$$
$$S = Ph + 2A$$
$$V = Ah$$

For prisms with regular polygons as bases, P = length of one side $\times$ number of sides. To obtain area of base, if it is a polygon, divide it into triangles and find sum of partial areas.

Frustum of Prism

If a section perpendicular to the edges is a triangle, square, parallelogram, or regular polygon,

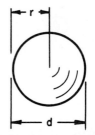

$$V = \frac{\text{sum of lengths of edges}}{\text{number of edges}} \times \text{area of right section}$$

Sphere

$$S = \pi d^2 = 4\pi r^2 = 12.5664 r^2$$
$$V = \frac{\pi d^3}{6} = \frac{4}{3}\pi r^3 = .5236 d^3 = 4.1888 r^3$$

Circular Ring

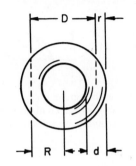

D = mean diameter
R = mean radius
$S = 4\pi^2 YRr = 9.8696Dd$
$V = 2\pi^2 Rr^2 = 2.4674Dd^2$

Wedge

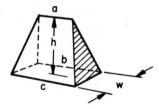

$$V = \frac{1}{6}wh(a + b + c)$$

3

Conversion Tables

It is often necessary to change weights, measures, and other factors from one system to another. For example, a dimension found on a working drawing may be given in millimeters, and it is desired to change the dimension from the SI system to the U.S. equivalent in inches, or fraction of an inch. To simplify this operation, the Tables 3.1 and 3.2 are designed to enable the user to quickly find the equivalent of practically any system that will be used in machine shop work.

Table 3.2 is designed so that one weight or measure is listed in the left-hand column. The mathematical operation is then given to convert this designation to the desired one, and the correct designation for the resulting figure can be found (in most cases) in the center column. The format of Table 3.1 should be self-explanatory.

TABLE 3.1 Temperature Conversion

Use this table to convert Fahrenheit degrees (F°) directly to Celsius degrees (C°) and vice versa. It covers the range of temperatures used in most hardening, tempering and annealing operations. Lower, higher and intermediate conversions can be made by substituting a known Fahrenheit (F°) or Celsius (C°) temperature figure in either of the following formulas:

$$F° = \frac{C° \times 9}{5} + 32 \qquad C° = \frac{F° - 32}{9} \times 5$$

F°	C°	F°	C°	F°	C°	F°	C°	F°	C°
−160	−107	340	171	840	449	1340	727	1840	1004
−140	− 96	360	182	860	460	1360	738	1860	1016
−120	− 84	380	193	880	471	1380	749	1880	1027
−100	− 73	400	204	900	482	1400	760	1900	1038
− 80	− 62	420	216	920	493	1420	771	1920	1049
− 60	− 51	440	227	940	504	1440	782	1940	1060
− 40	− 40	460	238	960	516	1460	793	1960	1071
− 20	− 29	480	249	980	527	1480	804	1980	1082
0	− 18	500	260	1000	538	1500	816	2000	1093
20	− 7	520	271	1020	549	1520	827	2020	1104
40	4	540	282	1040	560	1540	838	2040	1116
60	16	560	293	1060	571	1560	849	2060	1127
80	27	580	304	1080	582	1580	860	2080	1138
100	38	600	316	1100	593	1600	871	2100	1149
120	49	620	327	1120	604	1620	882	2120	1160
140	60	640	338	1140	616	1640	893	2140	1171
160	71	660	349	1160	627	1660	904	2160	1182
180	82	680	360	1180	638	1680	916	2180	1193
200	92	700	371	1200	649	1700	927	2200	1204
220	104	720	382	1220	660	1720	938	2220	1216
240	116	740	393	1240	671	1740	949	2240	1227
260	127	760	404	1260	682	1760	960	2260	1238
280	138	780	416	1280	693	1780	971	2280	1249
300	149	800	427	1300	704	1800	982	2300	1260
320	160	820	438	1320	716	1820	993	2320	1271

TABLE 3.2 General Conversion Factors

Multiply	By	To obtain
centimeters	0.3397	inches
centimeters	0.01	meters
centimeters	393.7	mils
centimeters	10	millimeters
centimeter-grams	980.7	centimeter-dynes
centimeter-grams	10^{-5}	meter-kilograms
centimeter-grams	7.233×10^{-5}	pound-feet
centimeters per second	1.969	feet per minute
centimeters per second	0.03281	feet per second
centimeters per second	0.036	kilometers per hour
centimeters per second	0.6	meters per minute
centimeters per second	0.02237	miles per hour
centimeters per second	3.728×10^{-4}	miles per minute
cubic centimeters	3.531×10^{-5}	cubic feet
cubic centimeters	6.102×10^{-2}	cubic inches
cubic centimeters	10^{-6}	cubic meters
cubic centimeters	1.308×10^{-6}	cubic yards
cubic centimeters	2.642×10^{-4}	gallons
cubic centimeters	10^{-3}	liters
cubic centimeters	2.113×10^{-3}	pints (liquid)
cubic centimeters	1.057×10^{-3}	quarts (liquid)
cubic feet	62.43	pounds of water
cubic feet	2.832×10^{4}	cubic centimeters
cubic feet	1728	cubic inches
cubic feet	0.02832	cubic meters
cubic feet	0.03704	cubic yards
cubic feet	7.481	gallons
cubic feet	28.32	liters
cubic feet	59.84	pints (liquid)
cubic feet	29.92	quarts (liquid)
cubic feet per minute	472.0	cubic centimeters per second
cubic feet per minute	0.1247	gallons per second
cubic feet per minute	0.4720	liters per second
cubic feet per minute	62.4	pounds of water per minute

TABLE 3.2   General Conversion Factors [*Continued*]

Multiply	By	To obtain
cubic inches	16.39	cubic centimeters
cubic inches	5.787×10^{-4}	cubic feet
cubic inches	1.639×10^{-5}	cubic meters
cubic inches	2.143×10^{-5}	cubic yards
cubic inches	4.329×10^{-3}	gallons
cubic inches	1.639×10^{-2}	liters
cubic inches	0.03463	pints (liquid)
cubic inches	0.01732	quarts (liquid)
cubic yards	7.646×10^{5}	cubic centimeters
cubic yards	27	cubic feet
cubic yards	46,656	cubic inches
cubic yards	0.7646	cubic meters
cubic yards	202.0	gallons
cubic yards	764.6	liters
cubic yards	1616	pints (liquid)
cubic yards	807.9	quarts (liquid)
degrees (angle)	60	minutes
degrees (angle)	0.01745	radians
degrees (angle)	3600	seconds
dynes	1.020×10^{-3}	grams
dynes	7.233×10^{-5}	pounds
dynes	2.248×10^{-6}	pounds
feet	30.48	centimeters
feet	12	inches
feet	0.3048	meters
feet	.36	varas
feet	⅓	yards
grains (troy)	1	grains (avoirdupois)
grains (troy)	0.06480	grams
grains (troy)	0.04167	pennyweights (troy)
grams	980.7	dynes
grams	15.43	grains (troy)
grams	10^{-3}	kilograms
grams	10^{3}	milligrams
grams	0.03527	ounces
grams	0.03215	ounces (troy)

TABLE 3.2 General Conversion Factors (*Continued*)

Multiply	By	To obtain
grams	0.07093	poundals
grams	2.205×10^{-3}	pounds
horsepower	42.44	British thermal units per minute
horsepower	33,000	foot-pounds per minute
horsepower	550	foot-pounds per second
horsepower	1.014	horsepower (metric)
horsepower	10.70	kilogram-calories per minute
horsepower	0.7457	kilowatts
horsepower	745.7	watts
inches	2.540	centimeters
inches	10^3	mils
inches	.03	varas
kilograms	980,665	dynes
kilograms	10^3	grams
kilograms	70.93	poundals
kilograms	2.2046	pounds
kilograms	1.102×10^{-3}	tons (short)
$\log_{10} N$	2.303	$\log_\epsilon N$ or $\ln N$
$\log_\epsilon N$ or $\ln N$	0.4343	$\log_{10} N$
meters	100	centimeters
meters	3.2808	feet
meters	39.37	inches
meters	10^{-3}	kilometers
meters	10^3	millimeters
meters	1.0936	yards
ounces	8	drams
ounces	437.5	grains
ounces	28.35	grams
ounces	.0625	pounds
ounces	0.0625	pounds per square inch
pounds	444,823	dynes
pounds	7000	grains
pounds	453.6	grams
pounds	16	ounces

TABLE 3.2   General Conversion Factors (*Continued*)

Multiply	By	To obtain
pounds	32.17	poundals
pounds per cubic foot	0.01602	grams per cubic cm.
pounds per cubic foot	16.02	kilograms per cubic meter
pounds per cubic foot	5.787×10^{-4}	pounds per cubic inch
pounds per cubic foot	5.456×10^{-9}	pounds per mil foot
square centimeters	1.973×10^{5}	circular mils
square centimeters	1.076×10^{-3}	square feet
square centimeters	0.1550	square inches
square centimeters	10^{-6}	square meters
square centimeters	100	square millimeters
square feet	1.296×10^{-5}	acres
square feet	929.0	square centimeters
square feet	144	square inches
square feet	0.09290	square meters
square feet	3.587×10^{-8}	square miles
square feet	.1296	square varas
square feet	⅑	square yards
square inches	1.273×10^{6}	circular mils
square inches	6.452	square centimeters
square inches	6.944×10^{-3}	square feet
square inches	10^{6}	square mils
square inches	645.2	square millimeters
temperature + 17.8	1.8	temperature (°F)
temperature − 32	⅝	temperature (°C)
tons (long)	2240	pounds
tons (short)	2000	pounds
yards	.9144	meters

SOURCE: Dietzen Corporation.

4

The Metric System*

The metric system is an international language of measurement. Its symbols are identical in all languages. Just as the English language is governed by rules of spelling, punctuation, and pronunciation, so is the language of measurement. Uniformity of usage facilitates comprehension, and minimizes the chance of decimal point errors.

This chapter covers the basic rules of the metric system, and is designed to serve as a convenient reference guide. It uses the modern form of the metric system, called "The International System of Units," which is abbreviated SI. A limited number of the units outside SI that are acceptable for use with SI are also included.

Except for the use of American instead of British spelling, this guide is consistent with the International Standard ISO 1000 and the National Bureau of Standards Special Publication 330, which

_____

* See the text on this page for the newer name used for this measurement system.

is an approved English-language translation of the International Bureau of Weights and Measures publication "Le Systéme International d'Unités." However, greater standardization of punctuation, and a selection of preferred prefixes, are recommended in this presentation, and the pronunciation of unit names is given.

RULES FOR WRITING METRIC QUANTITIES

1. *Capitals.*   Unit names, including prefixes, are not capitalized except at the beginning of a sentence or in titles. Note that in degree Celsius the unit "degree" is lowercase. (In the term "degree Celsius," "degree" is considered to be the unit name, modified by the adjective "Celsius," which is always capitalized. The "degree centigrade" is now obsolete.)

(a) *Symbols.*   The short forms for metric units are called symbols (see Table 4.1). They are lowercase except that the first letter is uppercase when the name of the unit derived from the name of a person, for example:

Unit name	*Unit symbol*
meter	m
liter	l
gram	g
newton	N
pascal	Pa

Printed unit symbols should have upright letters, because sloping (italic) letters are reserved for quantity symbols, such as for mass and length.

(b) *Prefix symbols.*   All prefix names, their symbols, and pronunciation are listed in Table 4.2. Notice that the top five

TABLE 4.1

Quantity (1)	Common units	Symbol	Acceptable equivalent	Symbol
Plane angle	degree (2)	°		
Length	kilometer	km		
	meter (3)	m		
	centimeter	cm		
	millimeter	mm		
	micrometer	μm		
Area	square kilometer	km^2		
	square hectometer	hm^2	hectare	ha
	square meter	m^2		
	square centimeter	cm^2		
	square millimeter	mm^2		
Volume	cubic meter	m^3		
	cubic decimeter	dm^3	liter (3,4,5)	l
	cubic centimeter	cm^3	milliliter (4)	ml
Velocity	meter per second	m/s		
	kilometer per hour	km/h		
Acceleration	meter per second squared	m/s^2		
Frequency	megahertz	MHz		
	kilohertz	kHz		
	hertz	Hz		
Rotational frequency	revolution per second	r/s		
	revolution per minute	r/min		rpm
Mass	megagram	Mg	metric ton	t
	kilogram	kg		
	gram	g		
	milligram	mg		

TABLE 4.1 (Continued)

Quantity (1)	Common units	Symbol	Acceptable Equivalent	Symbol
Density	kilogram per cubic meter	kg/m^3	gram per liter	g/l
Force	kilonewton	kN		
	newton	N		
Moment of force (6)	newton-meter	N·m		
Pressure	kilopascal (7)	kPa		
Stress	megapascal (8)	MPa		
Energy, work, or quantity of heat	megajoule	MJ		
	kilojoule	kJ		
	joule	J		
	kilowatt-hour (9)	kW·h	kilowatthour	kWh
Power or heat flow rate	kilowatt	kW		
	watt	W		
Temperature	kelvin	K		
	degree Celsius	°C		
Electric current	ampere	A		
Quantity of electricity	coulomb	C		
	ampere-hour (10)	A·h		^h
Electromotive force	volt	V		

TABLE 4.1 *(Continued)*

Quantity (1)	Common units	Symbol	Acceptable Equivalent	Symbol
Electric resistance	ohm	Ω		
Luminous intensity	candela	cd		
Luminous flux	lumen	lm		
Illuminance	lux	lx		
Sound level	decibel	dB		

(1) Listed in same sequence as ISO 1000.

(2) For efficiency in calculations, plane angles should be expressed with decimal subdivisions rather than minutes (′) and seconds (″). The SI unit for plane angle is radian (symbol rad), defined as "the plane angle between two radii of a circle which cut off on the circumference an arc equal in length to the radius."

(3) The spellings "meter" and "liter" are preferred but "metre" and "litre" are recognized to be in widespread use.

(4) To be used only for fluids (both gases and liquids) and for dry ingredients in recipes.

(5) Do not use any prefix with "liter" except "mili."

(6) Torque or bending moment.

(7) See rule 10.

(8) Except for very weak materials.

(9) To be abandoned eventually. 1 kW·h = 3.6 MJ

(10) 1 A·h = 3600 C

TABLE 4.2 SI Unit Prefixes

Multiplication factor	Prefix	Symbol
1 000 000 000 000 000 000 = 10^{18}	exa (2)	E
1 000 000 000 000 000 = 10^{15}	peta (2)	P
1 000 000 000 000 = 10^{12}	tera	T
1 000 000 000 = 10^{9}	giga	G
1 000 000 = 10^{6}	mega	M
1 000 = 10^{3}	kilo	k
100 = 10^{2}	hecto	h (4)
10 = 10	deka	da (4)
0.1 = 10^{-1}	deci	d (4)
0.01 = 10^{-2}	centi	c (4)
0.001 = 10^{-3}	milli	m
0.000 001 = 10^{-6}	micro	μ (5)
0.000 000 001 = 10^{-9}	nano	n
0.000 000 000 001 = 10^{-12}	pico	p
0.000 000 000 000 001 = 10^{-15}	femto	f
0.000 000 000 000 000 001 = 10^{-18}	atto	a

(1) The first syllable of every prefix is accented to assure that the prefix will retain its identity. Therefore, the preferred pronunciation of kilometer places the accent on the first syllable, not the second.

(2) Approved by the 15th General Conference of Weights and Measures (CGPM), May–June, 1975.

(3) These terms should be avoided in technical writing because the denominations above one million are different in most other countries, as indicated in the last column.

Pronunciation (U.S.A.) (1)	Meaning (in U.S.A.)	In other countries
ex′ a (*a* as in *a*bout)	One quintillion times (3)	trillion
		thousand
as in *petal*	One quadrillion times (3)	billion
as in *terrace*	One trillion times (3)	billion
jig′ a (*a* as in *a*bout)	One billion times (3)	milliard
as in *mega*phone	One million times	
as in *kilo*watt	One thousand times	
heck′ toe	One hundred times	
deck′ a (*a* as in *a*bout)	Ten times	
as in *deci*mal	One tenth of	
as in *senti*ment	One hundredth of	
as in *mili*tary	One thousandth of	
as in *micro*phone	One millionth of	
nan′ oh (*an* as in *an*t)	One billionth of (3)	milliardth
peek′ oh	One trillionth of (3)	billionth
		thousand
fem′ toe (*fem* as in *fem*inine)	One quadrillionth of (3)	billionth
as in ana*to*my	One quintillionth of (3)	trillionth

(4) While hecto, deka, deci, and centi are SI prefixes, their use should generally be avoided except for the SI unit-multiples for area and volume and nontechnical use of centimeter, as for body and clothing measurement. The prefix hecto should be avoided also because the longhand symbol h may be confused with k.

(5) Although rule 1 prescribes upright type, the sloping form is sometimes tolerated in the U.S.A. for the Greek letter μ because of the scarcity of the upright style.

are uppercase and all the rest lowercase. The importance of following the precise use of uppercase and lowercase letters is shown by the following examples:

G for giga; g for gram
K for kelvin; k for kilo
M for mega; m for milli
N for newton; n for nano
T for tera; t for metric ton

2. *Plurals and Fractions.*

(a) Names of units are plural when appropriate. EXAMPLES: 1 meter, 100 meters, 0 degrees Celsius. Values less than 1 take the singular form of the unit name. EXAMPLES: 0.5 kilogram or ½ kilogram. While decimal notation (0.5, 0.35, 6.87) is generally preferred, the most simple fractions are acceptable, such as those where the denominator is 2, 3, 4, or 5.

(b) Symbols of units are the same in singular and plural. EXAMPLES: 1 m, 100 m.

3. *Periods.*   A period is NOT used after a symbol, except at the end of a sentence. EXAMPLES: A current of 15 mA is found. . . . The field measured 350 × 125 m.

4. *The Decimal Marker.*   A dot on the line is used as the decimal marker. In numbers less than one, a zero should be written before the decimal point (to prevent the possibility that a faint decimal point will be overlooked). EXAMPLE: The oral expression "point seven five" is written 0.75.

5. *Grouping of Numbers.*

(a) Separate digits into groups of three, counting from the decimal marker. A comma should not be used. Instead, a space is left to avoid confusion, because many countries use a comma for the decimal marker.

(b) In a four-digit number, the space is not required unless the four-digit number is in a column with numbers of five digits or more.

EXAMPLES: For 4,720,525 write 4 720 525
            For 0.52875   write 0.528 75
            For 6,875   write 6875 or 6 875
            For 0.6875   write 0.6875 or 0.687 5

6. *Spacing.*

(a) In symbols or names for units having prefixes, no space is left between letters making up the symbol or the name. EXAMPLES: kA, kiloampere; mg, milligram.

(b) When a symbol follows a number to which it refers, a space must be left between the number and the symbol, except when the symbol (such as °) appears in the superscript position. The symbol for degree Celsius may be written either with or without a space before the degree symbol. EXAMPLES: 455 kHz, 22 mg, 20 mm, 10^6 N, 30°, 20°C or 20° C.

(c) When a quantity is used in an adjectival sense, a hyphen should be used between the number and the symbol (except ° and °C). EXAMPLES: It is a 35-mm film. . . . The film width is 35 mm. I bought a 6-kg turkey. . . . The turkey weighs 6 kg.

(d) Leave a space on each side of signs for multiplication, division, addition, and subtraction—except within a compound symbol. EXAMPLES: 4 m $\times$ 3 m (not 4 m$\times$3 m); kg/m^3; N$\cdot$m.

7. *Squares, Cubes, etc.*   When writing symbols for units such as square meter or cubic centimeter, write the symbol for the unit,

followed by the superscript ^{2} or ^{3}, respectively. EXAMPLES: For 14 square meters, write 14 m^2. For 26 cubic centimeters, write 26 cm^3.

8. *Compound Units.*

(a) For a unit name (not a symbol) derived as a quotient—for example, kilometers per hour—it is preferable not to use a slash (/) as a substitute for "per" except where space is limited and a symbol might not be understood. Avoid other mixtures of words or symbols. EXAMPLES: Use meter per second; not meter/second, meter/s, or m/second. Use only one "per" in any combination of units. EXAMPLE: meter per second squared, not meter per second per second.

(b) For a unit symbol derived as a quotient—for example, km/h—do not write k.p.h. or kph because these are understood only in the English language, whereas km/h is used in all languages. The symbol km/h can also be written by means of a negative exponent—for example, $km \cdot h^{-1}$. Do not use more than one slash (/) in any combination of symbols unless parentheses are used to avoid ambiguity. EXAMPLES: m/s^2, not m/s/s; W/(m$\cdot$K), not W/m/K.

(c) For a unit name derived as a product, a hyphen is recommended (or a space is permissible) but never a "product dot" (a period raised to a centered position)—for example, write newton-meter or newton meter, not newton$\cdot$meter.

(d) For a unit symbol derived as a product, use a product dot—for example, N$\cdot$m. Do not use the product dot as a multiplier symbol for calculations. EXAMPLE: Use 6.2 $\times$ 5, not 6.2$\cdot$5

(e) Do not mix nonmetric units with metric units, except units for time, plane angle, or rotation. EXAMPLE: Use kg/m^3, not kg/ft^3.

(f) To eliminate the problem of what units and multiples to use, a quantity that constitutes a ratio of two like quantities should be expressed as a fraction (either common or decimal) or as a percentage. EXAMPLES: The slope is ¹⁄₁₀₀ or 0.01 or 1 percent, not 10 mm/m or 10 m/km.

9. *Prefix Usage.*

(a) While hecto, deka, deci, and centi are SI prefixes, their use should generally be avoided exept for the SI unit-multiples for area and volume and nontechnical use of centimeter, as for body and clothing measurements.

(b) In broad fields of use, or in a table of values for the same quantity, or in a discussion of such values within a given context, a common unit-multiple should be used even when some of the numerical values may require up to five or six digits before the decimal point.

> EXAMPLES: mm   for mechanical engineering drawings
> kPa   for fluid pressure
> MPa   for stress
> kg/m^3 for density

(c) When convenient, choose prefixes resulting in numerical values between 0.1 and 1000, but only if this can be done without violating rule 9*a* or *b*.

(d) To avoid errors in calculations, prefixes may be replaced with powers of 10—for example, $1\ MJ = 10^6\ J$.

(e) Generally avoid use of prefixes in a denominator, except kilogram—for example kJ/m^3, not J/dm^3, and kJ/kg, not J/g.

(f) Do not use a mixture of prefixes, unless the difference in size is extreme. EXAMPLES: Use 40 mm wide and 1500 mm long, not 40 mm wide and 1.5 m long; but 1500 meters of 2-mm-diameter wire.

(g) Do not use multiple prefixes. EXAMPLES: Use 13.58 m, not 13 m 580 mm. Use nm, not mμm. Use milligram, not microkilogram.

(h) Do not use a prefix without a unit—for example, use kilogram, not kilo.

10. *Units for Pressure.*   Kilopascal (kPa) is the only unit recommended for fluid pressure for all fields of use except high-vacuum measurements of absolute pressure for which Pa, mPa, etc. may be more convenient. Do not use bar (10^5 Pa) or millibar (10^2 Pa) because they are not SI units, and are accepted internationally only for a limited time in special fields because of existing usage. They are also objectionable because they introduce too many different units, requiring frequent conversions to the preferred SI unit kPa (10^3 Pa), with consequent chance for decimal point errors. (1 bar = 100 kPa)

Gage pressure is absolute pressure minus ambient pressure (usually atmospheric pressure). It is positive or negative (called vacuum) according to whether the pressure is higher or lower, respectively, than the ambient pressure. Absolute pressure is specified either by (1) using the identifying phrase "absolute pressure," or (2) adding the word "absolute" after the unit symbol, separating the two by a comma or a space. Do not add (to the unit symbol) either "g" for gage or "a" for absolute.

11. *Spelling of Vowel Pairs.*   There are three cases where the final vowel in a prefix is omitted—namely, megohm, kilohm, and hectare. In all other cases, both vowels are retained and both are pronounced. No space or hyphen should be used.

PRONUNCIATION OF METRIC TERMS

The pronunciation of most of the unit names is well-known and uniformly described in American dictionaries, but four have been pronounced in various ways. The following pronunciations are recommended:

Candela. Put the accent on the second syllable and pronounce it like dell.

Joule. Pronounce it to rhyme with pool.

Pascal. The preferred pronunciation rhymes with rascal. An acceptable second choice puts the accent on the second syllable.

Siemens. Pronounce it like seamen's.

For pronunciation of unit prefixes, see Table 4.2, especially footnote 1.

WEIGHT, MASS, AND FORCE

Considerable confusion exists in the use of the term weight as a quantity to mean either force or mass. In commercial and everyday use, the term weight nearly always means mass; thus, when one speaks of a person's weight, the quantity referred to is mass. This nontechnical use of the term weight in everyday life will probably persist. On the other hand, in physics, weight has usually meant the force of gravity. American dictionaries define weight as either the heaviness or the mass of an object.

To avoid the ambiguity of this dual use of the term weight, it is better to avoid its use in technical practice except under circumstances in which its meaning is completely clear. Instead, the terms mass and force of gravity (or gravity force) should be used, together with the appropriate units kilogram (kg) and newton (N) respectively.

A floor load rating or a capacity rating of a vehicle (or other load-supporting machine) is intended to define the mass that can be supported safely. Hence the rating should be expressed in kilograms rather than newtons. Similarly, the area density of floor loading should be expressed in kilograms per square meter rather than pascals (newtons per square meter). In engineering calculations involving structures, vehicles, or machines on the surface of the earth, the mass in kilograms is multipled by 9.8 to obtain the approximate force of gravity in newtons. (The force of gravity acting on a mass of 1 kg varies from about 9.77 N to 9.83 N in various parts of the world.)

TYPEWRITING RECOMMENDATIONS

Superscripts

The question of how superscripts should be typed on a machine with a conventional keyboard is raised by rule 7.

With an ordinary keyboard, figures and the minus sign can be raised to the superscript position by rolling the platen half a space before typing the figure. When this is done, the figure may tend to run into the text in the line above. This interference can be avoided by using care. In printing, interference is avoided by making superscripts of smaller type than the body of the text, but this procedure would be available with a typewriter only by modifying the keyboard.

Special Characters

For technical work it is useful to have a number of Greek letters available on the typewriter. If all metric symbols for units are to be properly typed, a key with the Greek lowercase μ (pronounced "mew" like a cat, not "moo" like a cow) is necessary, since this is the symbol for "micro," meaning one millionth. This can be approximated on a conventional machine by using a lowercase μ, and adding the tail by hand (µ). While this is not very elegant, it is unmistakable, and those planning typed work can choose whether to use this procedure, or use the transfer characters known as press-ons, or spell the unit name out in full.

For units of electricity, the Greek uppercase omega (Ω) for ohm will also be useful, but when it is not available, the word "ohm" can be spelled out.

It should be noted that, except for the more extensive use of the Greek μ for "micro" and Ω for ohm, the change to metric measurements will cause no more difficulty than has already existed in typing of material containing quantitative information.

The Letter L for Liter

On most typewriters there is no difference at all between the lowercase letter "l" (the recognized symbol for liter) and the figure "one." Accordingly, it is preferable to spell the word in full—for example, 24 liters/100 km (fuel consumption). However, there is no problem with ml (milliliter).

In a situation where you cannot afford the space for the full word liter, and you do not want to use a lowercase symbol l because it looks like a number 1, a capital L (although not yet recognized in any American National Standard or International Standard) is preferable to a script $\mathcal{L}$, because it does not preempt a typewriter character.

A better symbol than l is badly needed, and is now being considered by national and international standards bodies.

Typewriter Modification

Where frequently used, the following symbols should be included on typewriters: superscripts ^{2} and ^{3} for squared and cubed; Greek

μ for micro; ° for degree; · for a product dot, for symbols derived as a product; Greek Ω for ohm.

The positions in which these are placed on the keyboard will depend partly on the design of the keyboard, and partly on the frequency with which the new characters are likely to be needed. A special type ball is available for some typewriters which contains all the superscripts, μ, Ω, and many other characters used in technical reports. Some typewriters have replaceable character keys.

LONGHAND

To assure legibility of the symbols m, n, and u, it is recommended that these three symbols be written to resemble printing. For example, write n m—not m n . The symbol μ should have a long distinct tail, ʍ.

SHORTHAND

Stenographers will find that the SI symbols are generally quicker to write than the shorthand forms for the unit names.

5

Machine-Shop Drawings

A blueprint or shop drawing is an exact copy or reproduction of an original drawing, consisting of lines, symbols, dimensions, and notations to accurately convey a design to the workers who produce the part in the machine shop. A blueprint, therefore, is an abbreviated language for conveying a large amount of exact, detailed information, which would otherwise take many pages of manuscript or hours of verbal instruction to convey.

In every branch of machine-shop work, there is often occasion to read a shop drawing. A machinist, for example, will consult the drawings to determine the type of material used in the object to be made. Dimensions are referred to, as well as other pertinent information.

A drawing for the machine trades ideally should be so informative and complete that the workers can take it and, without further verbal or written instructions, produce the object as the designer intended it to be made. The information contained in a machine-shop drawing should cover the form and size of the object, the kind of material from which it is to be made, the number of pieces desired, and the finish of its surface. Table 5.1 shows the ANSI-recommended line widths and conventions for engineering and machine-shop drawings.

TYPES OF DRAWINGS

Most machine-shop drawings will fall under one or more of the following categories: working drawings, assembly drawings, tool drawings, installation drawings, and manufacturing drawings. These drawings may also be classified as a pictorial drawing, orthographic projection drawing, a diagram, a section view, and the like.

PICTORIAL DRAWINGS

In pictorial drawings, an object is drawn in one view only. Three-dimensional effects are simulated on the flat plane of drawing paper by drawing several faces of an object in a single view. In most cases, these drawings are used to convey information to those people who are not well-trained in blueprint reading or to supplement the conventional orthographic drawings in more complex designs.

The pictorial drawings most often encountered in machine shop work include:

- Isometric drawing
- Oblique drawings
- Perspective drawing

TABLE 5.1 ANSI Lines for Engineering Drawings

 Cast and malleable form (Also for general use of all materials)

 Titanium an. refractory material

 Steel

 Electric windings, electromagnets, resistance, etc.

 Bronze, brass, copper and compositions

 Concrete

 White metal, zinc, lead, babbitt, and alloys

 Marble, slate, glass, porcelain, etc.

 Magnesium, aluminum, and aluminum alloys

 Earth

 Rubber, plastic electrical insulation

 Rock

 Cork, felt, fabric, leather, fiber

 Sand

 Sound insulation

 Water and other liquids

 Thermal insulation

 Wood—across grain
Wood—with grain

The main disadvantage of pictorial drawings is that intricate parts cannot be pictured clearly and most are difficult to dimension.

Orthographic Projection Drawings

These drawings are used more than any other in plans for homes and smaller projects. Orthographic projection drawings generally give all plan views, elevation views, dimensions, and other details necessary to construct a project or object.

To illustrate the practicality of the orthographic drawing, look at the pictorial drawing in Fig. 5.1. While this view clearly suggests the form of a block, it does not show the actual shape of the surfaces. Also, it does not show the dimensions of the object so that it may be constructed.

An orthographic projection of this same block appears in Fig. 5.2. The front view in this drawing shows the block as though it

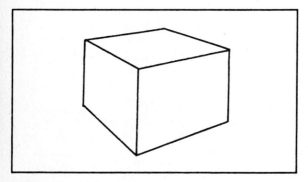

FIG. 5.1 Pictorial drawing of a cube.

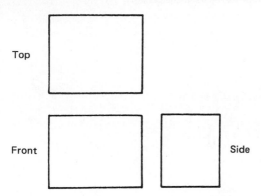

FIG. 5.2 Orthographic projection of the cube shown in Fig. 5.1

was viewed straight at the front. Another view is shown looking straight at the left side, while another is as viewed from the top. These views, when combined with dimensions, will allow the object to be constructed. The material may be specified in written specifications, a schedule, or by merely added a note to the drawing.

Diagrams

Diagrams are drawings that are intended to show components and their relationship in diagrammatic form. Such drawings are seldom drawn to scale and show only the working association of the different components. Symbols are used extensively in diagrams to represent the various pieces of equipment or components. (See Fig. 5.3)

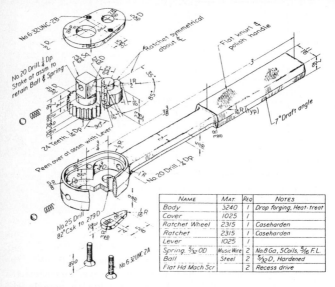

NAME	MAT.	REQ	NOTES
Body	3240	1	Drop forging, Heat-treat
Cover	1025	1	
Ratchet Wheel	2315	1	Caseharden
Ratchet	2315	1	Caseharden
Lever	1025	1	
Spring, 5/32 OD	Music Wire	2	No.8 Ga, 5 Coils, 5/16 F.L.
Ball	Steel	2	5/32 D, Hardened
Flat Hd Mach Scr		2	Recess drive

FIG. 5.3 A simple production diagram used to indicate order of assembly.

Sectional Views

A section of any object is what could be seen if the object was sliced or sawed into two parts at the point where the section is taken. If it is desired to see how a golf ball is constructed, for example, the ball could be secured in a vise and then cut with a hacksaw. When the two parts are separated, the interior of the ball may be seen.

Considerable visualization must be used when dealing with sections, as there are no given rules for determining what a section will look like. If a round hollow shaft, for example, is cut vertically, the section will appear as shown in Fig. 5.4. If cut horizontally, the section will appear as shown in Fig. 5.5. If cut on a slant, the section will form an ellipse as shown in Fig. 5.6

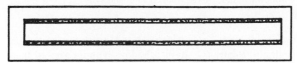

FIG. 5.4 A section taken vertically through a hollow round shaft.

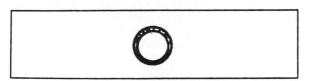

FIG. 5.5 A section cut horizontally through a hollow round shaft.

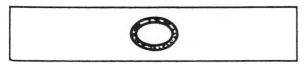

FIG. 5.6 A section through a hollow round shaft cut on a slant—forming an ellipse when drawn.

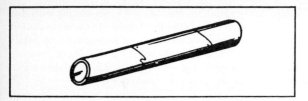

FIG. 5.7 Hollow cylinder viewed from one end and side.

The theory of construction for a sectional view is shown in Figs. 5.7 through 5.9. Figure 5.7 shows a hollow cylinder viewed from the end and side. In Fig. 5.8 a cutting plane is shown passing through the pipe, that is, where the section is taken and is indicated by the cutting plane line. The portion of the pipe section between the viewer and the cutting plane has been removed to reveal the interior details of the cylinder. In Fig. 5.9 the cutting plane is removed and the section would be drawn in an orthographic view.

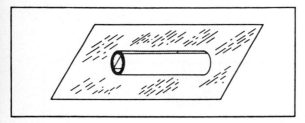

FIG. 5.8 An imaginary cutting plane passing through the cylinder in Fig. 5.7

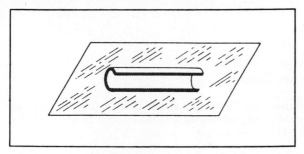

FIG. 5.9 The section nearest the viewer has been removed to reveal the interior of the cylinder.

Section lining or crosshatching is used in sectional views to indicate the various materials of construction. For common metal such as steel, lining is made with fine lines, usually drawn at angles of 45 degrees as shown in Fig. 5.10

Other sectional views are classified as:

- Full section
- Half section

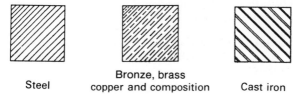

Steel

Bronze, brass
copper and composition

Cast iron

FIG. 5.10 Common section lining or crosshatching used in machine shop sectional views.

- Revolved section
- Removed section
- Aligned section
- Broken-out section
- Assembly section

Full section.   A full section is a view in which the cutting plane is assumed to pass entirely through the object. The sectional view in Fig. 5.11 is a full section.

Half section.   A half section is a sectional view in which the cutting plane passes halfway through the object. One half of the view is shown in section, while the other half is shown from the exterior. Figure 5.12 shows a cutting plane passing halfway through a steel shaft while Fig. 5.13 shows the section removed.

Revolved section.   A revolved section is a cross section that has been revolved through 90 degrees. It is used to show the true shape of the cross section of bars and other elongated parts. Figure 5.14 shows such a section.

Removed section.   A removed or detail section is a cross section that has been removed from its original position to a convenient space near the principal view. See Fig. 5.15.

Aligned section.   An aligned section is a sectional view in which a sloping part is rotated parallel to the cutting plane to show its true shape. See Fig. 5.16.

Broken-out section.   A broken-out section is used when less than half a section is sufficient to show some interior detail (Fig. 5.17).

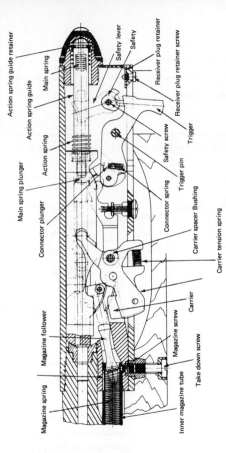

FIG. 5.11 A full section of a .22 caliber rim fire, semiautomatic rifle.

Action spring guide retainer

Safety lever

Safety

Receiver plug retainer

Receiver plug retainer screw

Action spring guide

Main spring

Trigger

Action spring

Safety screw

Trigger pin

Main spring plunger

Connector spring

Connector plunger

Carrier spacer Bushing

Carrier tension spring

Magazine follower

Magazine screw

Carrier

Magazine spring

Take down screw

Inner magazine tube

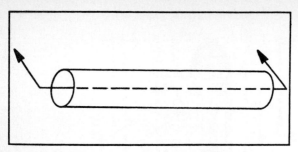

FIG. 5.12 Cutting plane passing halfway through a steel shaft.

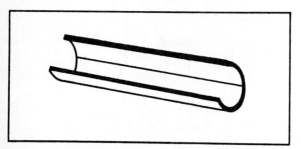

FIG. 5.13 Section of the steel shaft in Fig. 5.12 removed to reveal the interior of the object.

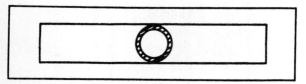

FIG. 5.14 A revolved section of the steel shaft in Fig. 5.12.

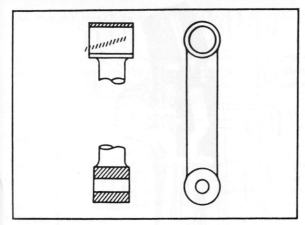

FIG. 5.15 A removed section.

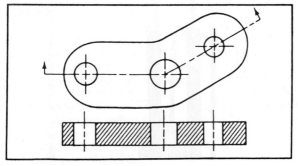

FIG. 5.16 An aligned section.

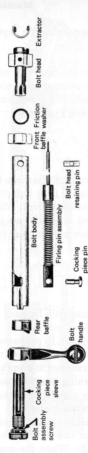

FIG. 5.18 An assembly section of a bolt for a Savage Model 110 centerfire rifle.

Extractor

Bolt head

Friction washer

Front baffle washer

Bolt body

Firing pin assembly

Bolt head retaining pin

Cocking piece pin

Rear baffle

Cocking piece sleeve

Bolt handle

Bolt assembly screw

indicate the type of metal by symbols, notes are normally provided near the items to indicate the type of metal to use. A few symbols for metals are shown in Fig. 5.19.

DRAWING DIMENSIONS

A drawing is expected to convey exact information regarding every detail so that an object can be fabricated. It would be impossible to achieve successful results without definite dimensions on the drawings.

A drawing dimension is a numerical value expressed in appropriate units—feet and inches, metric units, and the like. They are indicated on drawings in conjunction with lines, symbols, and notes to define the geometrical characteristics of an object.

Dimensions are usually shown between points, lines, or surfaces that have a necessary and specific relation to each other or that control the location of other components or mating parts. See Fig. 5.20. Furthermore, dimensions should be shown only once on a drawing (so as not to confuse). Only enough measurements are given so that the intended sizes, shapes, and locations can be determined without assuming any distances.

APPLICATIONS

One type of important drawing for the machine shop trades is that pertaining to jigs and fixtures. Jigs are used in performing identical operations easily and rapidly with uniformity of precision. Jigs are specifically adapted for such operations as drilling, reaming, counterboring, and tapping. A jig is used to hold the workpiece in position and to guide the cutting tool.

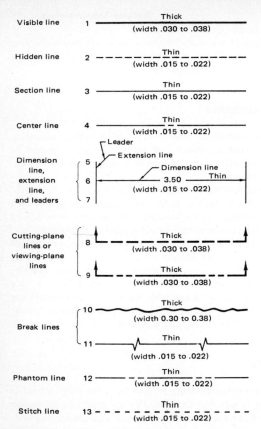

Visible line	1	Thick (width .030 to .038)
Hidden line	2	Thin (width .015 to .022)
Section line	3	Thin (width .015 to .022)
Center line	4	Thin (width .015 to .022)
Dimension line, extension line, and leaders	5 6 7	Leader Extension line Dimension line 3.50 Thin (width .015 to .022)
Cutting-plane lines or viewing-plane lines	8 9	Thick (width .030 to .038) Thick (width .030 to .038)
Break lines	10 11	Thick (width 0.30 to 0.38) Thin (width .015 to .022)
Phantom line	12	Thin (width .015 to .022)
Stitch line	13	Thin (width .015 to .022)

FIG. 5.19 Symbols used for metal in machine-shop drawings.

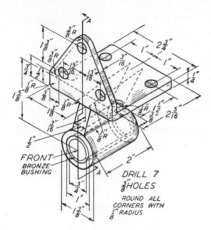

FIG. 5.20 Typical dimensions on a machine-shop drawing.

Fixtures are also locating and holding devices. Unlike jigs, they are clamped in a fixed position and are not free to move or guide the cutting tool. Most fixtures are used to perform operations requiring facing, boring, milling, grinding, welding, and the like.

The following are recommended by Brown and Sharpe Manufacturing Company for designing and producing jigs and fixtures:

- Before laying out the jig or fixture, decide on the locating points and outline a clamping arrangement.

- Make all clamping and binding devices as quick-acting as possible. These devices may be purchased from various sources and should be used whenever possible.

- See that two component parts of a machine can be located from corresponding points and surfaces.
- Make the jig foolproof. Arrange it so that the work cannot be inserted except in the correct way.
- Make some of the locating points adjustable for rough castings.
- Locate clamps so that they will be in the best position to resist the pressure of the cutting tool during the operation.
- If possible, make all clamps integral parts of the jig or fixture.
- Avoid complicated clamping arrangements that may wear or get out of order.
- As nearly as possible, place all clamps opposite from the bearing points of the work to avoid springing.
- To make the tools as light as possible, core out all unnecessary metal.
- Round all corners.
- Provide handles wherever they will make the handling of the jig more convenient.
- Provide feet (preferably four) opposite all surfaces containing guide bushings in drilling and boring jigs.
- Place all bushings inside the geometrical figure formed by connecting the points of location of the feet.
- Provide sufficient clearance, particularly for wrought castings.
- If possible, make all locating points visible to the operator when placing the work in position.
- Provide holes or escapes for the chips.
- Locate clamping lugs to prevent springing of fixtures on all tools that must be held to the table of the machine. Provide

tongues that will fit slots in the machine tables in all milling and planing fixtures.

- Use ASA drill jig bushings—namely, head press-fit bushings and slip-renewable bushings on all new tool designs.
- Use liners where slip-renewable bushings are used.
- Used headless liners for all general applications.

Shop drawings of simple devices are often confined to a single assembly that incorporates the essential details. For more complicated devices, both an assembly drawing and parts detail drawings are prepared.

Assembly Drawings

Most assembly drawings are prepared with sufficient views so that the assembly is clearly understood. Sectional views are often necessary to illustrate the construction as discussed previously. Figure 5.21 is a typical assembly drawing and details of a tap wrench. The very top view in the upper left-hand corner is the side view of the wrench and the one immediately under it is a plan view. It is drawn as though the wrench is lying flat on a table and the viewer is looking straight down on it. Other details of the drawing include a section of the body, spring, and adjusting sleeve.

The sectional view of the body indicates that the wrench is made of steel, along with the adjusting sleeve. When the plan view is examined, it will be noted that the plunger fits in the end of the body. The spring fits between this plunger and the wrench body. Notes also indicate that the spring is to be made from no. 14, .032-in gage music wire. Because the adjusting sleeve is threaded, it may be adjusted by turning it either clockwise or counterclockwise.

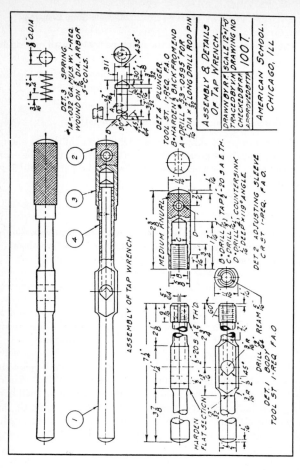

FIG. 5.21 Machine-shop drawing of a tap wrench.

The project in Fig. 5.21 has few parts and assembly of this wrench may be done without using reference numbers. However, if many parts were required in the assembly, it would be compulsory to use reference numbers on the drawing to facilitate reading it. Reference numbers are, however, included in this drawing to give an idea of how they are arranged on a working drawing. For example, reference numbers 1, 2, 3, and 4 on the drawing are referred to in the notes as body, adjusting sleeve, spring, and plunger, respectively.

Letters are also used in this drawing to convey information. On the plan view the dimensions are not specified for the tapped and drilled holes. The tapped hole is specified by B, and the drilled hole is indicated by c. Notes on drawings give necessary details.

The object in Fig. 5.22a is shown in an isometric view. The mechanical drawing in Fig. 5.22b requires two views to fully describe the object. Two solid lines are used to represent the projection on the block in the front view, and the actual shape of the projection is shown in the plan or top view.

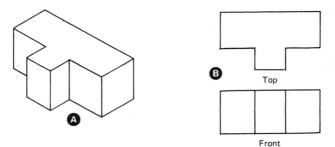

FIG. 5.22 Isometric view of a block along with a plan and front view.

Figure 5.23a shows an isometric view of a v block. The vertical lines in the front view represent the edges of the v block's various parts. A top view is required to complete all the facts regarding these lines. The middle line in the front view represents the corner in the bottom of the v. Again, two views are required to relate all the facts concerning the object.

WORKING DRAWINGS

Working drawings usually indicate all the necessary details for the construction of a single or multiple object—not the assembly of the parts, but rather how each part of the final object is to be constructed. Figure 5.24, for example, shows a working drawing for the construction of a brass ring that is 3 in in diameter and ⅛ in thick. The thickness of the ring is given on the end view. The large hole in the ring is 2 in in diameter. This dimension is found

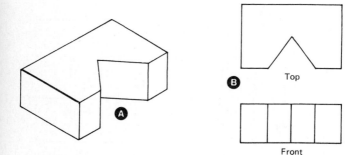

FIG. 5.23 Various views of a 'v' block.

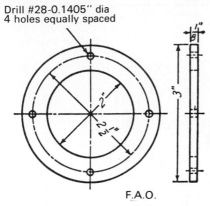

Drill #28-0.1405″ dia
4 holes equally spaced

F.A.O.

FIG. 5.24 Working drawing of a brass ring.

on the front view. There are four small holes drilled through the ring that are equally spaced. The size of these holes is indicated directly above the drawing as drill no. 28, .1405 in in diameter. These holes are located on a circle that is 2½ in in diameter. The end view indicates that these holes are drilled completely through the ring due to the dotted lines being shown in the drawing.

The letters F.A.O. indicate that this piece is machined or finished all over.

A perspective view of this ring is shown in Fig. 5.25. An experienced craftsperson does not need this sketch to construct the object. However, the sketch does enable the beginner to better visualize the finished object's appearance.

A more complicated machine shop drawing is in Fig. 5.26 which shows a wheel spindle head used on a tool grinding machine. An

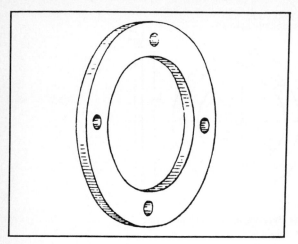

FIG. 5.25 Perspective view of the brass ring shown in Fig. 5.24.

assembly drawing for this same object is shown in Fig. 5.27. Note that the assembly drawing consists of two views—a front elevation and a right end elevation. In the former view the upper part is shown in section, the central part is a full view, and the lower part is partly indicated. In the right-end elevation the central part is fully shown, and the lower part is omitted. Practices vary in making assembly drawings. Some assembly drawings have all the piece parts indicated by numbers, thereby aiding the mechanic in assembling the pieces. Piece parts or detail numbers are not specified in this case.

At the top of the front elevation or center part of the machine, some of the pieces are enclosed. The piece that covers them is the

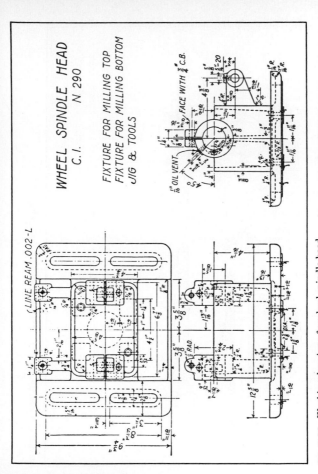

FIG. 5.26 Working drawing of a wheel spindle head.

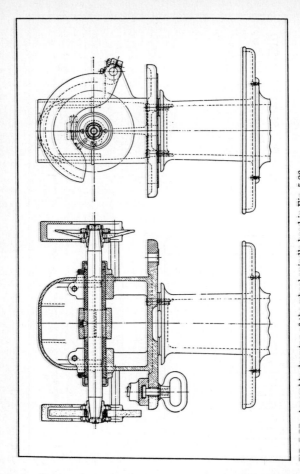

FIG. 5.27 Assembly drawing of the wheel spindle head in Fig. 5.26.

wheel spindle head. A spindle rotates within the wheel spindle head, which has bearing housings on each end with bushings inside the housings. This spindle holds the emery wheels. A pulley is located inside the wheel spindle head and between the bearings; as can be seen in the sectional view, the pulley is made of cast iron and held to the spindle by a headless screw. Two pieces are shown in the section to the right and left of the pulley. These pieces are babbitted bushings that aid in keeping down friction. Caps are shown at the extreme left and right of the babbitted bushings. The caps come in contact with the right and left bearings. These caps are fastened to the bushings with headless setscrews.

In the right end elevation, the part of the head that is attached to the pedestal is grooved out somewhat at the top. This is indicated by invisible or dotted lines on the drawing.

To the right and left of the machine's centerline are lines tapering up from the bottom of the head to the bottom of the bearing. These lines indicate the sides of the ribs. Refer to the sectional part of the front elevation. The perpendicular lines directly below the bearings and set back slightly from the bearings' inner ends indicate the length of the ribs.

The upper part of the bearing housings are not crosshatched in the front sectional elevation. Refer to the right end elevation at the upper part of the housings. The front sectional elevation was obtained by taking a cutting plane on the centerline in the right end elevation. Here two invisible perpendicular lines may be seen, one to the right and one to the left, extending from the top of the bearing down to the bushings. You will also find to the right and to the left of the top part of the housings a screw passing through, which indicates that the housings are split at the top.

In the front elevation at the extreme right end of the spindle, a certain piece directly above the sectional elevation is very vague in this particular view. However, in referring to the right elevation, it can be seen that the piece is a guard used to cover the top of the grinding wheel. The guard keeps flying pieces of a broken wheel from hitting the worker. The connection of this guard to the machine may be found by referring to the right end elevation. Visible and invisible lines may be seen connecting to a circular part. The lines indicate that a bracket connects the guard to the wheel spindle head. On the end of the arm is a boss with a hole through it which is fitted with a pin. The pin allows the guard to turn when it is necessary to remove the wheel from the spindle. It is necessary to raise the guard out of the way to remove the wheel from the spindle.

A flat fillister head screw is shown in the front elevation view on the right side and at the top of the pedestal. Two fillister head screws are shown in the right-end elevation. This indicates that the wheel spindle head is fastened to the pedestal with four fillister head screws.

Continuing with the drawing at hand, the specifications for the wheel spindle head may be found on the right of the plan view; C.I. indicates that cast iron is the material, and N 290 is the pattern number. The notes refer to special tools available for mass production, Fig. 5.26.

The width of the spindle head is given at 9¾ in on the left of the plan view, but the length is not shown in this view. The length dimension, 12⅜ in, can be found by referring to the front elevation.

In studying the front elevation and focusing on the bottom part, a horizontal invisible line can be found located ⁷⁄₁₆ in from the bottom horizontal line. This shows that the base has an impression in it, but the length and width of the impression are not given. The person building the object will make the top part of the base 2⅜ in on each end and then allow for the length of the impression. The impression starts a short distance from the back and front side and terminates at the perpendicular line. The extreme width of the box-like design on the casting is shown. The correct location cannot be obtained from this particular view. A ¼-in thickness is specified in the plan view.

Figure 5.26 shows that the housings are supported by ribs directly below them. To determine the distances from the outer edge of the vertical part of the casting to the edge of the rib, refer to Fig. 5.26 between the right housing and the base. The distance is 1⁵⁄₁₆ in. On some blueprints this dimension would be specified on the right side elevation; on others it would be on the front elevation.

The housings are located on the base. Refer to the front elevation to the right of the perpendicular centerline, and note that the dimension is 1¹³⁄₁₆ in. This is the distance from the centerline to the inside face of the bearing housing. This dimension is used for locating the other housing. For the length of the housing, add 1¹³⁄₁₆ to 1⅝ in; the result is 3⁷⁄₁₆ in. Subtract 3⁷⁄₁₆ in from 3⅝ in. The difference is ³⁄₁₆ in. Add ³⁄₁₆ to 1⅝ in. The total is 1¹³⁄₁₆ in. To the right of the left housing is the dimension 2¾ in, which is the outside diameter of the housing. The dimension of the diameter of the hole in the housing is 1⅝ in.

After looking at the front elevation of the working drawing, it is difficult to determine whether parts of the housings are enclosed in a box-like casting. Study the right-hand elevation. Note that the bearings are partly encased in a rectangular box-like casting. Refer to the plan view. The corners of this box are rounded, and the radius of these corners must be found before they can be properly constructed. Below the right housing in the front right-hand corner of the box, the radius is shown as ⅜ in, which is the radius to use for the inner corner. The outside radius, as seen from the rear right-hand corner of the box, is ⅝ in.

Lugs that are to be machined through the center with a ⅛-in saw can be found on the right end elevation directly above the housing. Their design or shape is not indicated. Refer to the front elevation and study the various parts of the bearing housing. The housing to the left shows that the ear or lug has a radius of ½ in. Not knowing just how high the top part of this lug is to come, again read the blueprint. The distance from the center of the housing to the center of the hole is not specified on the front elevation. Refer to the right end elevation to find this dimension, which is %₁₆ in. This gives the correct position to set dividers when scribing the arc that will give rounding for the ear.

To find how the grinding wheel guard is supported, refer to the drawings and note that the guard swivels on a pin inserted in a hole in a bracket that is at the back of the box-like part of the casting. The location of the bracket relative to the top and center of the box may be found by locating the horizontal centerline. On the right end elevation above the bracket and the box-like part of the casting is the dimension 4⅝ in. This is the specification for the horizontal distance. Between the bearing and the box's right wall is the dimension 1½ inches—the distance from the top of the box to the center of the bracket in which the hole will be machined.

6

Measuring Tools and Measurements

The basic measuring tools consist of a good machinist's scale, outside and inside calipers, and the hermaphrodite caliper. With only these simple tools an experienced machinist can take measurements to a surprisingly high degree of accuracy. This is accomplished by developing a sensitive "feel" for the instruments and by carefully setting the calipers to that they "split the line" graduated on the scale.

The accepted method for setting an outside caliper to a steel scale is shown in Fig. 6.1. Note that the scale is held in the left hand with the caliper in the right. One leg of the caliper is held against the end of the scale and is supported by the index finger of the left hand while the adjustment is made with the thumb and first finger of the right hand.

When measuring the diameter of a cylinder, the outside caliper is held exactly at right angles to the centerline of the work and is pushed gently back and forth across the diameter of the cylinder to be measured. When the caliper is adjusted properly, it should

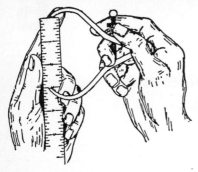

FIG. 6.1 Method of setting an outside caliper to a steel scale.

easily slip over the shaft of its own weight. The caliper should never be forced or it will spring, causing any further measurements to be inaccurate.

Inside calipers are adjusted in a similar way, except that the steel scale should be placed against a flat surface with one leg of the caliper resting squarely on the flat surface at the end of the scale. The other end of the caliper may then be adjusted to the required dimension.

To measure an inside diameter, place the inside caliper in the hole with the caliper ring spring slightly lower than the axis of the bore. Then raise the hand slowly until the vertex of the caliper legs is in line with the hole. Adjust the caliper until it will slip into the hole with a very slight drag, holding the caliper square across the diameter of the hole.

When taking measurements, it is sometimes desirable to transfer measurements from an outside caliper to an inside caliper. To do so, rest the point of one leg of the inside caliper on a similar point of the outside caliper. Using this contact point as a pivot, move the other inside caliper leg along the edge of the other outside caliper leg, and adjust with the thumb screw until the measurement is just right.

STEEL RULE

Most modern machinist's steel rules are made of tempered stainless steel with a satin chrome finish. Some of the types available are illustrated in Fig. 6.2.

On the same order as steel rules are steel scales. They are used mainly for measuring machine shop drawings—which are often drawn to either a reduced or larger scale than the object to be made actually is—that is, many machine shop drawings are not drawn to actual size.

The common method of reducing all the dimensions in the same proportion is to choose a certain distance and let that distance represent one foot; this distance is then divided into 12 parts and each one of these parts represents an inch; then if half and quarter inches are required, these twelfths are subdivided into halves, quarters, etc., until the subdivisions become so small that they cannot be used. We now have a scale which represents the common foot rule with its subdivisions into inches and fractions of an inch, but this "foot" is smaller than the ordinary distance, and of course its subdivisions are proportionately smaller. When a distance is laid off on a drawing, the scale is used; when measurements are made on the object to be made, a conventional true-size measuring device is used. Scales are also made in metric units as well as feet and inches.

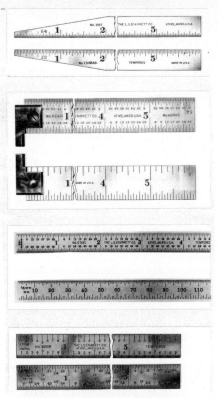

FIG. 6.2 Several types of steel rules available for use by the machinist. *(The L. S. Starrett Company.)*

Sometimes if the piece to be drawn is too small to be satisfactorily shown full size, the drawing is made to an enlarged scale, such as twice size, three times size, etc.

HERMAPHRODITE CALIPER

The hermaphrodite caliper shown in Fig. 6.3 is set from the end of the scale exactly the same as the outside caliper. This instrument is used in layout work for locating and testing centers, laying off distances from an edge, and the like.

On round work, such as cylinders, centers are usually located

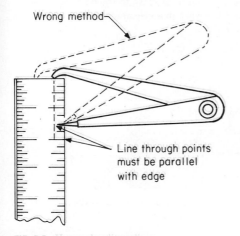

Wrong method—

Line through points must be parallel with edge

FIG. 6.3 Hermaphrodite caliper.

with either the hermaphrodite caliper or the center head attachment for a steel scale. In centering square, hexagonal, and other regular-sided stock, lines are scribed across the ends from corner to corner, and then center-punched at the point of intersection.

In using the hermaphrodite caliper, set the caliper to a little more than half the diameter of the work and scribe four lines. Then hold the work in a vise and center-punch as accurately as possible in the center of these marks. A little chalk rubbed over the end of the work before scribing enables the marks to be easily seen.

When the center head is used, set the center head as shown in Fig. 6.4 and scribe two lines approximately at right angles. Use a sharp scriber and keep the lines as close to the edge of the scale as possible. Then hold the work in a vise and center-punch at the intersection of the two lines.

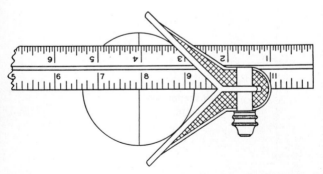

FIG. 6.4 Setting the center head to obtain the center of an object.

MICROMETER

Micrometers are available in many different types. Besides the ultra-accurate conventional measuring instruments, many professionals are using the electronic types that offer the advantage of direct reading of each measurement, plus an automatic conversion button to change from English to metrics. One such instrument is the Fowler Ultra-Cal electronic calipers as shown in Fig. 6.5. This instrument reads to .0005 in or .01 mm easily, even in the dim light which ensures accuracy. The Inch/Metric reading feature permits immediate change to the standard preferred by the user. The floating zero feature allows one to "zero" instantly

FIG.   6.5 Fowler   Ultra-Cal Electronic   Calipers.   *(Fowler Co.)*

at any position, permitting the reading of deviations from the nominal size without performing any math.

With the conventional micrometer, each graduation on the micrometer barrel (D in Fig. 6.6) represents one turn of the spindle or .025 in. Every fourth graduation is numbered and the figures represent tenths of an inch such as 4 × .025 = .100 in or ⅒ in.

The thimble (E) has 25 graduations, each of which represents one-thousandth of an inch. Every fifth graduation is numbered, from 5 to 25 on most micrometers.

Micrometers for measuring in the metric system are graduated to read in hundredths of a millimeter as shown in Fig. 6.7. For each complete revolution the spindle travels ½ or .50 mm, and two complete revolutions are required for 1.00 mm. Each of the upper set of graduations on the barrel represents 1 mm. Two revolutions of the spindle and every fifth graduation is numbered to read 0, 5, 10, 15, etc. The lower set of graduations subdivides each millimeter division into two parts.

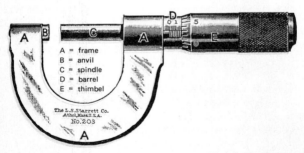

A = frame
B = anvil
C = spindle
D = barrel
E = thimbel

The L.S.Starrett Co.
Athol,Mass.U.S.A.
No.203

FIG. 6.6 Parts of a typical micrometer. *(The L. S. Starrett Company.)*

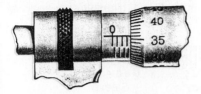

FIG. 6.7 Scales on a metric micrometer.
(The L. S. Starrett Company.)

The beveled edge of the thimble is divided into 50 graduations, each of which represents .01 mm.

The micrometer reading is the sum of the readings on the barrel and the thimble. For example, in Fig. 6.7, there are three millimeter graduations visible on the barrel, also a ½ mm graduation. The reading on the thimble is 36 mm. Therefore, the reading is 3.0 mm + .50 mm + .36 mm = 3.86 mm.

Measurements to ten-thousandths of an inch may be obtained with a vernier, an instrument invented by Pierre Vernier in 1631. Figure 6.8 shows the features of this type micrometer. When used with a micrometer, a vernier has 10 equal divisions on the sleeve, which are equal in length to exactly nine divisions on the thimble. The difference between the smallest division of the sleeve and the smallest division on the thimble is therefore 1/10 of a division on the thimble. Since a division on the thimble is equal to 1/1000 in, the difference will be one-tenth of this or 1/10,000 in.

Referring to Fig. 6.8*b*, note that the zero line of the thimble coincides with the zero on the sleeve, also the zero on the vernier is in line with one of the divisions of the thimble, indicating that the micrometer is open to a certain number of thousandths, but not ten-thousandths. However, in Fig. 6.8*c*, the zero of the thim-

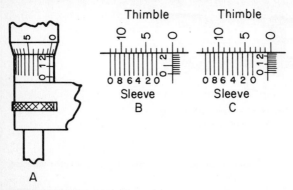

FIG. 6.8 Vernier micrometer.

ble is not in line with any division on the sleeve. The micrometer was opened to a certain number of thousandths, and a certain number of ten-thousandths. Looking at the vernier on the sleeve, note that line 7 coincides with one division on the thimble, indicating the number of ten-thousandths is 7. To read the micrometer opened as shown in Fig. 6.8c, first take the number of thousandths, which in this case is 0.250. To this is added 7 ten-thousandths (0.0007), and the reading will be 0.250 plus 0.0007 or 0.2507 in.

When measuring to ten-thousandths, use the vernier-type micrometer. First determine the number of thousandths the micrometer is open, and then, by looking at the vernier scale, determine which line coincides with a line on the thimble and indicates the number of ten-thousandths which are to be added to the previous reading.

Micrometers are available in a wide range of sizes from 0 to 1 in up to 39 to 40 in, or more.

The micrometers described thus far are called "outside micrometers", but several other types are available:

Point micrometers (Fig. 6.9). These are used to measure small grooves, keyways, and other hard-to-reach features. The measuring points on most models are .012 in and are available with or without carbide tips. The carbide increases life of the conical measuring points and improves accuracy.

Tubing-type micrometers (Fig. 6.10). These micrometers are highly useful and versatile tools. Primarily, they measure wall thicknesses of tubing and other parts with cylindrical walls. Dimensions between a hole and an edge of other shaped pieces can also be measured with these instruments.

Screw thread micrometers (Fig. 6.11). The screw thread micrometer with interchangeable anvils measures a pitch diameter. Each

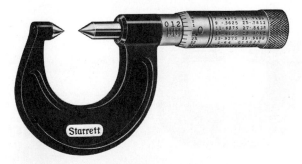

FIG. 6.9 Point micrometer. *(The L. S. Starrett Company.)*

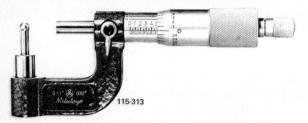

FIG. 6.10 Tubing-type micrometer.

anvil can be used for both metric and unified external threads. The correct pair of anvils must be selected and inserted in the hollow micrometer spindle and anvil.

Blade micrometers (Fig. 6.12). These instruments are used to measure narrow grooves, keyways, and other hard-to-reach locations.

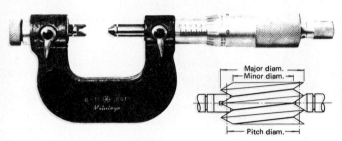

FIG. 6.11 Screw thread micrometer.

FIG. 6.12 Blade micrometer. *(The L. S. Starrett Company.)*

Hub micrometers (Fig. 6.13). This model incorporates an extra-shallow, specially designed frame to clear through a ¾-in-diameter hole and is designed for hub length measurements, but can be used in other areas where the regular c-shaped micrometer cannot be applied.

Uni-Mike (Fig. 6.14). This instrument has a vise-type rugged frame which holds two furnished interchangeable anvils and additional accessories. It is used with a selected anvil for most measuring applications, but is also used without anvils to measure a step or height from a flat surface as illustrated in Fig. 6.15.

FIG. 6.13 Hub micrometer. *(The L. S. Starrett Company.)*

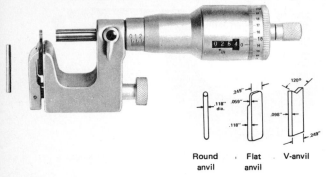

Round anvil   Flat anvil   V-anvil

FIG. 6.14   Uni-Mike.

Inside micrometer (Fig. 6.16).   The rod-type inside micrometer is used to check internal dimensions. It can also be used as an adjustable linear standard bar within the range.

Depth micrometer (Fig. 6.17).   The depth micrometer measures depth or step on work and is normally used with interchangeable rods.

There are also many other types of micrometers for special uses and with special shaped anvils. One of the tool catalogs should be consulted for an exact description of each model.

ANGLES AND ANGULAR MEASUREMENTS

When the lines of a workpiece are not parallel, it is either tapered (when round) or angular. Typical angular lines are shown in Fig. 6.18.

FIG. 6.15 Measuring a step or height from a flat surface.

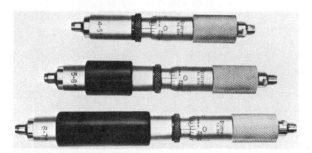

FIG. 6.16 Inside micrometer. *(The L. S. Starrett Company.)*

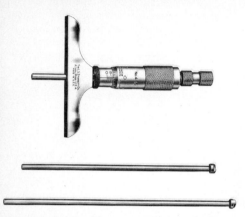

FIG. 6.17 Depth micrometer. *(The L. S. Starrett Company)*

Angles are usually dimensioned in degrees and minutes; a degree is a unit of angular measurement, and is equal to $\frac{1}{360}$ of a circle. Each degree in turn may be divided into minutes, and each minute into seconds. There are 60 minutes in a degree, and each minute has 60 seconds. In Fig. 6.18a there are three angles designated by A, B, and C. Angles can be measured either with a fixed angular gage or with a bevel protractor, shown in Fig. 6.19. This latter instrument is well-adapted to measurement of angles because the dial is graduated in degrees over the entire circle, and the vernier adds to the accuracy of reading, making possible measurement to at least 5 minutes, which is $\frac{1}{12}$ of a degree.

Each space on the vernier shown in Fig. 6.19 is 5 minutes shorter than two spaces on the scale. When the zero mark on the vernier coincides directly with zero graduation of the disc, the reading is obviously zero. When the zero mark on the vernier coincides directly with any division on the disc, the reading is in degrees, and when the zero graduation of the vernier does not coincide with a graduation on the scale, the reading will be in degrees and minutes.

Readings on the protractor can be taken in either direction

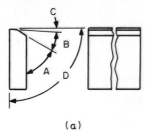

(a)

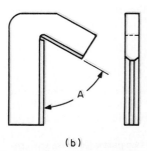

(b)

FIG. 6.18 Typical angles found in the machine shop.

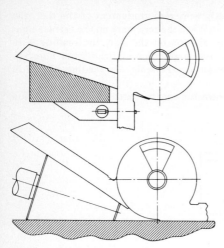

FIG. 6.19 Bevel protractor.

from zero, depending on which direction the zero on the vernier is moved.

VERNIER HEIGHT GAGE

The vernier height gage (Fig. 6.20) has many uses in the machine shop for measuring and marking off vertical dimensions. When equipped with a dial gage, it gives a sensitive "touch" for all types of layout work.

In general, a height gage consists of two parts: the base with the

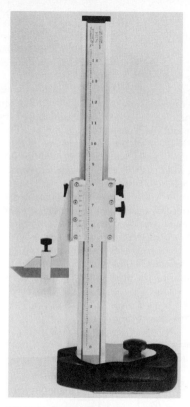

FIG. 6.20 Vernier height gage. *(The L. S. Starrett Company.)*

scale, and the slide or vernier. The slide is shown in Fig. 6.21. The main scale is divided into inches, and each inch is subdivided into 10 equal parts or divisions, and each of these ¹⁄₁₀ divisions is subdivided into four equal parts—making the smallest division on the scale ¹⁄₄₀ or 0.025 in.

The sliding or vernier scale has 25 divisions, numbered from 0 to 25. The 25 divisions on the vernier scale are equal to 24 divisions on the main scale—making the difference between divisions on the vernier and main scales ¹⁄₂₅ of ¹⁄₄₀, or ¹⁄₁₀₀₀ in.

If the vernier is so set that the zero line of the scale coincides with the zero line on the main scale, the reading is taken from the scale only. However, if the vernier scale is moved so that its zero line falls between any two lines on the main scale, then a certain line of the vernier will lie directly opposite a division of the main scale, indicating the number of thousandths to be added to the reading obtained from the main scale. For example, if, say, the fifth line of the vernier coincides with the line on the main scale, the reading of the vernier indicates ⁵⁄₁₀₀₀ in.

To read the vernier height gage, first note the number of inches between the end of the scale and the zero on the vernier scale;

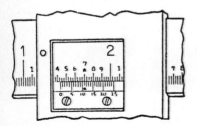

FIG. 6.21 Slide of a vernier height gage.

next note the number of tenths between the last whole inch and the zero of the vernier; then the number of ¹⁄₄₀ or 0.025 in divisions between the last ¹⁄₁₀ of the scale and the zero on the vernier, and finally look along the vernier scale and find the line that coincides with the line on the main scale, indicating the number of ¹⁄₂₀₀₀ parts of an inch. Sum up all readings.

In Fig. 6.21 the zero of the vernier lies between the first and second inch on the scale, showing that the height gage is opened one whole inch plus a fraction. Also, the zero of the vernier scale is past the line marked 4, indicating a further opening of ⁴⁄₁₀ inch. Note also that zero of the vernier is past the ¹⁄₄₀ or .025 inch line, and finally note that line 8 on the vernier coincides with a division on the scale, which means that .008 of an inch is added to the previous reading to obtain 1 + .4 + .025 + .008 = 1.433 inch— the amount the height gage is opened.

DIAL INDICATORS

To standardize minimum essential requirements for precision dial indicators, AGD specifications were introduced in 1945 from the National Bureau of Standards, as a Commercial Standard. See Fig. 6.22. They provide standards of performance and basic exterior dimensions in order to permit interchangeability.

1. *Repeatability.*   Readings shall agree within plus or minus one-fifth graduation at any point.

2. *Accuracy.*   Readings shall be accurate to within one graduation, plus or minus, at any point (2⅛ turns).

3. *Spindle range.*   The range of spindle travel should be a minimum of 2½ revolutions of the indicating hand.

FIG. 6.22 Typical dial indicator. *(The L. S. Starrett Company.)*

4. *Hand position.* The indicating hand shall be set at the approximate 9 o'clock position when the spindle is fuilly extended.

5. *Smallest graduation.* English: .00005 in, .0001 in, .0005 in, .001 in. Metric: .001 mm, .002 mm, .005 mm, and .01 mm.

6. *Dual numbering.* English: dial numbers shall always indicate thousandths of an inch regardless of magnification. Metric: dial numbers shall always indicate hundredths of a millimeter.

Metric and English test indicators are available in four types: horizontal, vertical, parallel, and extra long point. Most are normally used to check concentricity to a close tolerance as it is mounted on a machine.

ANGLE GAGE BLOCKS

A precision angle has always been difficult to set because of the involved trigonometric formula that is used with the sine bar.

The chief difficulty lies in the dimension X in Fig. 6.23a, which often results in a figure with many decimal places. Gage blocks can only approximate this value. For example, to measure 44 degrees, 30 minutes using a 5-in sine bar, the following steps are required:

Sine for 44 degree, 30 minute angle:	.7009093
For dimension X multiply by 5:	3.5045465
Gage blocks necessary to match this dimension:	.1005
	.104
	.300
	3.000
	3.5045
$3.5045465 - 3.5045 =$ residual error	.0000465

This error cannot be eliminated in the sine bar procedure.

With angle gage blocks you merely take a 45 degree block from the set, wring on a 30 minute block so that the plus end of a 45 degree block contacts the minus end of the 30 minute block and the result is an angle of 44 degrees, 30 minutes.

The diagram in Fig. 6.23b shows two 5 degree blocks wrung together with the plus and minus ends adjacent, resulting in two parallel lines.

A Webber angle block or true square is positioned on the work, and a beam of light from an autocollimator (Fig. 6.24) is directed against the gaging surface. This becomes 0 degrees, or the reference surface. Other angle blocks are then added in proper combination to measure each succeeding angle. The table is rotated and inspected at each position with reference to the light beam.

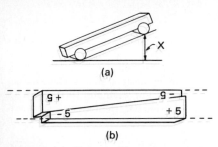

FIG. 6.23 (*a*) Example illustrating a typical machine-shop angle problem. (*b*) Two angle blocks in place, resulting in two parallel lines.

This method indexes large workpieces quickly with accuracy measured in fractional seconds of arc. See Fig. 6.25.

Webber angle gage blocks can be used with the work in the machine or in final inspection. The blocks reduce setup time and minimize error in grinding simple and compound angles. A workpiece on which an angle of 13 degrees is desired is placed on a parallel which is wrung to the angle blocks to form 13 degrees. The entire setup is lined up vertically with an angle plate, then indicated across the top of the work to determine the correctness of the angle.

Take a workpiece with a compound angle—the first angle 14 degrees, 30 minutes and the second angle 8 degrees, running at right angles to the first. To check the 14 degree, 30 minute angle,

FIG. 6.24 Use of autocollinotor in checking flatness or parallelism.

FIG. 6.25 A beam of light directed against the gaging surface for testing.

the work is laid on a parallel which is wrung to the proper combination of three blocks forming an angle of 14 degrees, 30 minutes. Correctness of this angle is then easily determined by using the dial indicator across the top surface of the work.

If the angle has been found correct, the inspection is continued by wringing together the proper blocks to form 8 degrees. These are laid on the 8 degree surface of the workpiece. After squaring them and setting them at right angles to the first surface, the correctness of the 8 degree angle can then be readily determined by using the dial indicator along the length of the gage blocks. The work can be placed on the surface plate or on top of the blocks, whichever is more convenient.

The method of setting a sine bar is as follows: To set the bar at an angle to the surface plate, the plugs are raised upon two piles of gage blocks so selected that when their difference in height is divided by the distance between the centers of the plugs, the result is the sine of the angle. Placing the cylinders 5, 10, or 20 in apart facilitates computation.

MEASURING THREADS

The pitch of a thread is defined as the distance from the crest of one thread to the crest of the next thread. The pitch of a threaded part can be readily determined by matching it against one of the toothed blades of a screw pitch gage, as shown in Fig. 6.26. These gages contain several different blades for a number of threads—usually ranging from 9 to 40. These gages are not intended as standards for thread profile or for use in determining errors in pitch, but only as a means of identifying the number of threads per inch and pitch.

When the pitch of the thread is to be measured to precise dimensions within limits of ten thousandths of an inch, other methods are used. One employs the projection comparator, in which the threads are projected on a screen of the projector in magnified form, with magnification of 5×, 10×, 20×, 50×, or 100×, and the pitch is then checked on the micrometer dial of the projector after the crest of one thread of the screw has been moved to the next crest against a fixed line on the projector screen. One type of projection comparator is shown in Fig. 6.27.

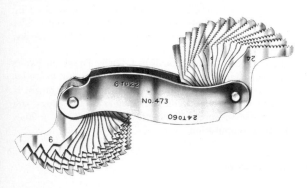

FIG. 6.26 Screw pitch gage. *(The L. S. Starrett Company.)*

FIG. 6.27 Projection comparator.

The thread profile (the exact shape of the thread) can also be checked with the projection comparator.

The lead of thread is the advance of the thread in one complete turn, or it is the advance of the nut along the screw in one complete turn. In a single-threaded screw the pitch and the lead are are equal; therefore, when the pitch of the threaded part is inspected, the lead is assumed to be the same.

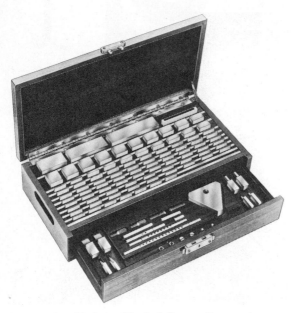

FIG. 6.28 Gage block set. *(The L. S. Starrett Company.)*

Lead of thread gages can best be checked for accuracy on a projection comparator or a toolmaker's microscope. When neither is available, precision gage blocks together with conical points and a holder can be used for checking the total lead. This method, however, depends very much on visual perception and is not as accurate as the comparator method.

GAGE BLOCKS

Gage blocks are often used to check various jigs, and usually a large assortment is necessary. Gage blocks are available in either English or metric, individually or in sets. See Fig. 6.28. All are made of a special steel alloy which possesses excellent thermal-expansion stability for most measuring applications.

Gage blocks are used for tool inspection and for reference purposes, or for setting up comparators for production inspection. After insertion of plugs into place, other measuring instruments are used to check the dimensions.

7

Machine-Shop Materials

Steel is the most frequently worked material in the machine shop. However, some plastic and fiber materials are also worked by machine tools.

Today, there are over 50 different kinds of steel in use. Some give greater strength for equal weight. Others will not corrode on exposure to acids or the elements, and others offer good wearing capabilities, after being heat-treated (Chap. 9), even after long use. Other steels are relatively soft and easily worked in forming dies to take unusual shapes.

Fortunately it is not necessary to learn all about the dozens of varieties of steel to meet the demands of the practical shop; a very few types will suffice for nearly every need.

Iron—the ancestor of all steel—is mined, and is originally in the state of ore before being refined to pig iron. Steel is produced from iron when carbon is added in such a way that it becomes part of the metal structure rather than an accretion of cinder or slag. The addition of carbon alone however, is not the sole distinction.

Cast iron, for example, contains approximately 3 percent to 5½ percent carbon in the form of uncombined graphite flakes, while in steel all of the carbon is normally combined with other elements, and the carbon seldom exceeds 2.2 percent.

Iron cannot be hardened, but steel can, and the amount of chemically combined carbon that is added determines how much hardness can be produced by proper heat treatment.

True cast iron is usually made by melting pig iron with scrapped castings; it has a higher carbon content than steel and, although it is hard, it is not malleable at any temperature. It is shaped by being heated to a liquid state and poured into a mold. Further heat treatment will produce a kind of hybrid cast iron just as strong as the original but sufficiently malleable to be bent and twisted without breaking.

Ordinary steel is a malleable union of carbon and iron. Its hardness in the annealed state can be increased by heating it to redness and quenching it, by plunging it into water or oil. The greater the carbon percentage, the greater the strength of the steel and the harder the steel can be made. The carbon content of steel is expressed in "points," each point being equal to .01 percent.

Low-carbon steel is chiefly employed in making shafts, structural members and other items in which hardness is not essential. But low-carbon steel can be surface-hardened, or case-hardened. Case-hardened steel, having a hard surface over a soft but tough core, is well-suited to the manufacture of dies, gears, engine parts, and similar articles that must be resilient enough to resist breakage and yet be able to withstand much surface abuse without wearing away or losing shape.

Most steel is an alloy steel, since common iron alone is too soft for most applications. Alloy steel is a steel to which other elements have been added. Carbon and manganese are the major alloys, but the percentage of each must be carefully regulated so that the steel will not be too brittle. The addition of nickel toughens steel, as does chromium. This latter additive is the main element that produces "stainless" steel. The addition of molybdenum increases the ability of steel to withstand shock and fatigue. Tungsten-alloy steel becomes stronger under heat and is the type most used in cutting tools such as drill bits, reamers, broaches, form cutters, and milling cutters.

Proper and exact identification of steel is almost impossible outside the laboratory, but reasonably accurate answers may be obtained in the shop by the spark test. With this method, a piece of the steel in question is held against the grinding wheel and the color, volume, and shape of the resulting sparks are noted. In general, a high-carbon steel produces a generous shower of many-branched white sparks. Low-carbon or mill steel produces light yellow sparks, but not in as great abundance and not as branched as high-carbon steel. High-speed steels produce relatively few sparks; such as they are, they are usually deep yellow with a tendency toward red—showing few, if any, branches.

A pocket magnet is another instrument to help identify certain metals. For example, the magnet will attract high-speed steels while it will not attract alloys composed of such materials as tungsten carbide. Stainless steels, however, will vary considerably; some types are magnetic, some are not, and grinding-wheel patterns will vary considerably. To test for a metal's "stainless" characteristics, apply a drop of vinegar to a polished surface. Allow the vinegar to dry and then rinse the metal with hot water. If no stain is left, the steel can be considered reasonably stainless.

Steel may currently be classified into five categories, for all practical purposes: (1) carbon, (2) alloy, (3) high-strength, low-alloy, (4) stainless, and (5) tool-and-die. Standard and tentative-standard steels in the first four categories above are now recognized under AISI-SAE numbers. A system of symbols is used to identify the grades of standard steels. Numbers are used to indicate grades of steel.

CARBON STEELS

High-carbon steel, also called ordnance steel, is usually composed of .45 to .55 percent carbon, with 1 to 1.3 percent manganese, and not more than .05 percent phosphorus, and not more than .05 percent sulfur. Also, steel known as SAE 1350, with .25 percent silicon added, is frequently used. Such steels have relatively high tensile strength, wear well, and are easily worked in the machine shop.

Manganese, next to carbon, is the most important ingredient in this type of steel, but too great a percentage of manganese makes the steel difficult to work, and too much carbon makes it too brittle.

The temperatures used for annealing, normalizing, or hardening carbon steels are as follows:

Up to .20 percent carbon: 1600 to 1650°F
From .20 to .35 percent carbon: 1550 to 1600°F
From .35 to .50 perent carbon: 1500 to 1550°F
From .50 to .70 percent carbon: 1450 to 1500°F
From .70 to .90 percent carbon: 1400 to 1450°F
With .90 percent carbon or over: 1350 to 1400°F

Carbon steels should be brought up to heat slowly in most cases, and after annealing, should be cooled slowly.

ALLOY STEELS

Steel is considered to be alloy steel when other elements, such as manganese, silicon, copper, etc. are added—in certain percentage limits—to steel. Of the alloying elements, manganese contributes to strength and hardenability; silicon increases the resiliency of steel for spring applications and raises the critical temperature for heat treatment; aluminum promotes nitriding properties; nickel lowers the critical temperature of steel and widens the temperature range for successful heat-treatment. Nickel is also used to promote resistance to corrosion.

Chromium is used mainly to promote hardness and promote the formation of carbides. Molybdenum increases hardenability, high-temperature tensile and creep strength. Vanadium promotes a fine austenitic grain size. Titanium acts as a deoxidizer in certain steels. Boron is another element added to increase hardenability.

A combination of two or more of the above elements usually imparts some of the characteristic properties of each.

HIGH-STRENGTH, LOW-ALLOY STEELS

This type of steel contains chemical compositions specially developed to impart improved mechanical properties and greater resistance to atmospheric corrosion.

STAINLESS STEELS

These steels fall into various categories including ferritic steels, which are nonhardenable, martensitic steels, which can be heat-treated, and austenitic steels, which are work-hardenable. All three types contain chromium to promote corrosiveness.

TOOL-AND-DIE STEELS

Tool-and-die steels fall into several categories for many purposes, the five principal ones being:

1. Heat resistance
2. Abrasion resistance
3. Shock resistance
4. Resistance to movement or distortion in hardening
5. Cutting ability

No one steel can develop these properties to the maximum extent, so hundreds of steels have been developed to meet practically every possible need. When another need arises, another type of steel will be developed. In general, however, tool-and-die steels will be either water-hardening, oil-hardening, or air-hardening. These steels are further developed into shock-resistant, high-speed, etc. types, with various alloys to obtain the desired characteristics.

One type of water-hardening tool steel drill rod, for example, consists of 1.00 percent carbon, .40 percent manganese, and .20 percent silicon and can be used for tools, parts, dies, or punches which require great hardness and a tough inner core for maximum strength. To harden this type of steel, heat the part to 1450 to 1500°F; hold at this point for 3 or 4 seconds and then quench in an 8 percent brine solution; that is, ¾ pound of rock salt to 1 gallon of water. Temper immediately. A 1 hour draw at the following temperatures will produce hardness on the Rockwell C scale as follows:

Temperature, °F	Rockwell C hardness
200	66–67
300	64–65
400	61–62
500	58–59
600	54–55
700	50–51
800	46–47

For spring temper, the above steel can be drawn at 650 to 750°F for 15 minutes.

One type of oil-hardening tool-steel drill rod consists of .95 percent carbon, 1.20 percent manganese, and .25 percent silicon. The characteristics of this steel include good machinability and smooth finish as well as superior dimensional stability and excellent hardening characteristics. This steel should be heated slowly to 1475 to 1525°F. Hold at this temperature for a few seconds and then quench in a light quenching oil. Temper immediately. The following hardness table is for 1-hour draw after oil quenching.

Temperature, °F	Rockwell C hardness
300	63–65
400	61–62
500	58–60
600	54–56
700	51–53
800	46–48
900	43–45

TABLE 7.1 Color Ranges and Tempering Treatments in Fahrenheit and Celsius Thermometer Standards.

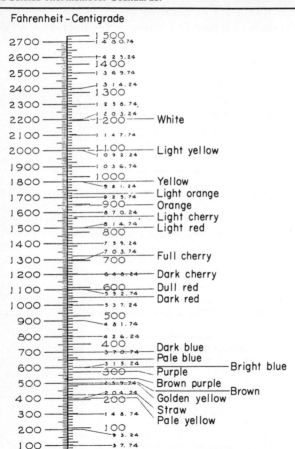

Table 7.1 shows the color ranges and treatments for carbon steel. Colors vary slightly with steel analysis and length of time held at a certain temperature.

Annealing temperatures and melting points of various metals are given in Table 7.2.

TABLE 7.2

Annealing temperatures		Metal melting points	
Metal	°F	Metal	°F
Aluminum	650	Pewter	425–440
Britannia metal	None	Brass	930
Pewter	None	Copper	2000
Brass	800–1300	Bronze	1825
Bronze	800–1200	Sterling silver	1640
Copper	700–1200	Fine silver	1760
Gold		Platinum	3223
Red	1100–1200	24 kt. gold	1945
White and yellow	1400	14 kt. yellow gold	1615
Nickel silver	1000–1400	14 kt. white gold	1825
Silver	900–1200	10 kt. yellow gold	1665
Stainless steel	1200	10 kt. white gold	1975

8

Soldering, Brazing, and Welding

Soldering, brazing, and welding are permanent assembly methods used in the machine shop. All three methods involve the use of heat to join metals. However, the amount of heat and the exact technique used with each varies considerably. In general, soldering is the method of joining metals by means of lead and tin mixed in a given proportion. Brazing involves a higher heat and the use of different materials. Welding is the most difficult of the three and requires more practice to master.

SOLDERING

Solders used to join various metals are composed of two or more metals. The solder itself must have a lower melting point than the metals that are being joined. Solders, in general, can be classified into two categories: hard and soft. Hard solder fuses at a red heat while soft solder requires a comparatively low temperature. Solders are further subdivided into a variety of classes such as brass, gold, copper, tin, plumber's solder, and the like.

Before soldering, the surfaces to be joined must be properly cleaned. Then the right flux must be selected. Still another requirement is the proper fitting of the edges of surfaces to be joined. The more accurate the fitting, the stronger the joint. Always use a solder with as high a melting point as possible, then apply proper heat. The nearer the temperature of the work to be joined is brought to the fusing point of the solder, the better will be the joint, since the solder will then flow more readily.

Fluxes

The purpose of a flux in soldering is to remove and prevent the formation of an oxide during the soldering operation and to allow the solder to flow readily and to unite more firmly with the surfaces to be joined. An excellent job of soldering can be done, on certain jobs that will permit it, by carefully fitting the parts, laying a piece of tin foil covered on both sides with a flux between the parts, then clamping them together, and heating the foil until it melts. This method is excellent on broken parts of brass and bronze.

Soft Solders

Soft solders consist mainly of lead and tin; other metals are sometimes added to lower the melting point. Lead-tin alloys will melt at a lower temperature as the percentage of tin is increased. However, when the percentage of tin exceeds 67 percent, the melting point rises gradually to the melting point of lead. The following table gives the melting temperatures of various lead-tin mixtures.

Tin %	Lead %	Melting temperature, °F
0	100	618
10	90	577
20	80	532
30	70	491
40	60	446
50	50	401

BRAZING (HARD SOLDERING)

Hard solder is used more frequently in the machine shop than soft solder. The operations of hard soldering and brazing are identical, and the two terms are often used interchangeably. There is a distinction, however. Brazing is the term used when joining metals with a film of brass, whereas hard soldering ordinarily means that silver solder is used as the uniting medium; the method is often termed "silver soldering."

For silver soldering or brazing, a red heat is normally required and a substance similar to borax is used as a flux to protect the metal from oxidization and to dissolve the oxides formed. As brazing and silver soldering require more heat than soft soldering, brazed work will withstand more heat without breaking or weakening and also has superior strength.

When brazing or silver soldering, the parts must first be thoroughly cleaned and then the parts placed in the position that is required. Usually the pieces are secured by clamps or soldering fixtures, but sometimes wire is used for holding the parts together.

The alloys used for brazing are composed of copper and zinc, and the melting point of these alloys depends upon the percentage of zinc. As the proportion of zinc increases, the melting point is lowered—usually from a high of around 2000°F to a low of 1300°F.

Silver solder contains silver, copper, and zinc or brass, the exact proportions varying according to the work it is to be used on. This is a very useful joining medium for repairing small parts, as the resulting joint will be extremely strong. Melting temperatures for silver solder run from a low of around 1000°F to around 1600°F. The lower temperature can be used on some heat-treated steels without affecting the hardness to any great extent.

WELDING

Welding cannot be learned by reading alone; skill comes only with practice. However, the following gives the basic principles.

Four simple factors are of prime importance in welding:

1. The correct welding position
2. The correct striking of the arc
3. The correct arc length
4. The correct welding speed

For arc welding, hold the electrode holder in the right hand (for right-handed persons), and then touch the left hand to the underside of the right hand for support. Place the left elbow into the left side. Whenever possible, welding should be done with two hands. This gives complete control over the movements of the electrode. Also, whenever possible, the right-handed person should weld from left to right to clearly see the operation. The electrode should be held at a slight angle as shown in Fig. 8.1.

Before striking an arc, make certain that the work clamp makes

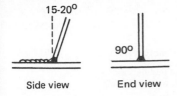

good electrical contact to the work. Then lower the head shield and scratch the electrode slowly over the metal, and sparks can be seen flying. While scratching, lift the electrode about ⅛ in and the arc should be established. However, the electrode should not be stopped during the scratching possess as it will stick to the work. This scratching should be done relatively slowly, as fast jabbing of the electrode will cause it to stick or else the arc will break immediately.

The arc length is the distance from the tip of the electrode core wire to the base metal. Once the arc has been established, maintaining the correct arc length becomes extremely important. The arc should be short, approximately ¹⁄₁₆ to ⅛ in long. As the electrode burns off, it must be fed closer to the work to maintain the correct arc length.

The easiest way to determine if the arc has the correct length is by listening to its sound. A nice, short arc has a distinctive, "crackling" sound, very much like eggs frying in a pan of grease. The inncorrect long arc has a hollow, blowing or hissing sound.

When welding, the puddle of molten metal, directly behind the arc, should be watched constantly. However, the arc itself should not be watched. The appearance of this puddle (Fig. 8.2) and the ridge where the molten puddle solidifies indicates correct welding speed. The ridge should be approximately ⅜ in behind the electrode. If the speed is too fast, the bead will result in a thin,

Ridge where puddle
solidifies

Molten puddle

FIG. 8.2 The ridge where the molten puddle solidifies should be approximately ⅜ in behind the welding electrode. *(Lincoln Electric Co.)*

uneven, "wormy" appearance. In general, thin material requires a faster speed while heavier plates will require a slower speed (moving of the electrode across the work).

Metals for Welding

Many types of metals encountered will be of the low-carbon type, which is made into items such as sheet metal, plate, pipe, channels, angle irons, and I beams. This type of steel can usually be easily welded without special precautions. Some steel, however, contains higher carbon including steel used in axles, connecting rods, shafts, scraper blades, and the like. These higher-carbon steels can be welded successfully in most cases, but care must be taken to follow proper procedures including preheating the metal to be welded and, in some cases, carefully controlling the temperature during and after the welding process. Regardless of the type of metal being welded, it is important, for quality weld, that the metal of free of oil, paint, rust or other contaminants.

Types of Welds

The five most common welded joints are butt welds, fillet welds, lap welds, edge welds, and corner welds. Of these, the butt and fillet welds are the most common. See Fig. 8.3.

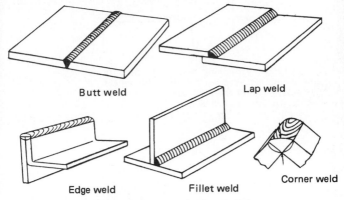

FIG. 8.3 Various types of welds. *(Lincoln Electric Co.)*

Butt welds. Butt welds are the most widely used welds. To make a butt weld, place two plates side by side, leaving about ⅟₁₆ in space between them for thin materials and ⅛ in for heavy metals. This distance allows adequate penetration of the weld. Start by tacking the plates at both ends as shown in Fig. 8.4a, otherwise the heat will cause the plates to move apart as shown in Fig. 8.4b.

The plates may now be welded together. In most cases, the weld should run from left to right (if you are right-handed). Point the electrode down in the crack between the two plates, keeping the electrode slightly tilted in the direction of travel. See Fig. 8.5. Watch the molten metal to be sure it distributes itself evenly on both edges and between the plates.

Unless a weld penetrates close to 100 percent, a butt weld will be weaker than the material welded together. In Fig. 8.6, for example, the total weld is only half the thickness of the material, making the weld only about half as strong as the metal.

FIG. 8.4 When performing butt welds, tack the plates at both ends as shown in *a*; starting at one end and working toward the other end will cause the plates to separate. *(Lincoln Electric Co.)*

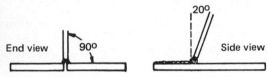

FIG. 8.5 Correct position of electrode to work for butt welding. *(Lincoln Electric Co.)*

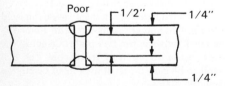

FIG. 8.6 In this weld the total weld is only one-half the thickness of the material—making the weld approximately half as strong as the metal. *(Lincoln Electric Co.)*

In the example in Fig. 8.7, the joint has been flame-beveled or ground prior to welding so that 100 percent penetration could be achieved. The weld, if properly made, is as strong or stronger than

Good

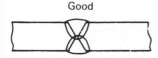

FIG. 8.7 This is a good welding joint giving 100 percent penetration. *(Lincoln Electric Co.)*

the original metal. When heavier metals are used, successive passes must be used to build up butt welds as shown in Fig. 8.8.

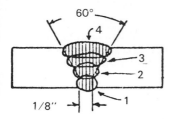

FIG. 8.8 Successive passes must be used to build up butt welds on heavier metal. *(Lincoln Electric Co.)*

Fillet welds. When welding fillet welds, it is important to hold the electrode at a 45 degree angle between the two sides as shown in Fig. 8.9, or the metal will not distribute itself evenly. To facil-

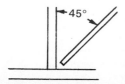

FIG. 8.9 When welding fillet welds, it is important to hold the electrode in a 45 degree angle between the two sides to distribute the weld evenly. *(Lincoln Electric Co.)*

itate this angle, the electrode is placed in the holder at a 45 degree angle as shown in Fig. 8.10.

Multiple passes are recommended for fillet welds as shown in

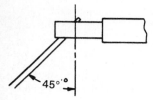

FIG. 8.10 To insure the correct welding angle for fillet welds, place the electrode in the holder at a 45 degree angle. *(Lincoln Electric Co.)*

Fig. 8.11. Put the first bead in the corner with fairly high current. Hold the electrode angle needed to deposit the filler beads as shown, putting the final bead against the vertical plate.

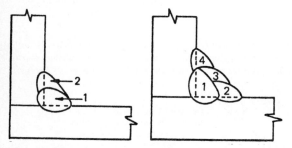

FIG. 8.11 Method of multiple pass welds. *(Lincoln Electric Co.)*

Vertical Welding

Welding in the vertical position can be done either vertical-up or vertical-down. Vertical-up is used whenever a large, strong weld is desired, while vertical-down is used primarily on sheet metal for fast, low-penetrating welds.

Vertical-up welding. The problem with this type of welding is to put the molten metal where it is wanted and make it stay there. If too much molten metal is deposited, gravity will pull it downward and make it "drip." Therefore, a certain technique has to be followed to ensure good results.

1. Use a ⅛-in (90 to 105 amps) or ³⁄₃₂-in (60 amps) Fleetweld 180 electrode.

2. When welding, keep the electrode horizontal or pointing slightly upward as shown in Fig. 8.12.

3. Strike the arc and deposit metal at the bottom of the two pieces to be welded together.

4. Before too mcuh molten metal is deposited, move the arc slowly ½ to ¾ in upward. This takes the heat away from the molten puddle, which solidifies. If the arc is not taken away soon enough, too much metal will be deposited, and it will drip.

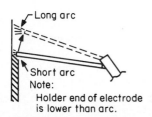

Long arc

Short arc

Note:
   Holder end of electrode
   is lower than arc.

FIG. 8.12 Demonstration of vertical-up welding. Note long and short arcs. *(Lincoln Electric Co.)*

5. The upward motion of the arc is caused by a very slight wrist motion. Most definitely, the arm must not move in and out, as this makes the entire process very complicated and difficult to learn.

6. If the upward motion of the arc is done correctly with a wrist motion, the arc will automatically become a long arc that deposits little or no metal. (Again see Fig. 8.12.)

7. During this entire process the only thing to watch is the molten metal. As soon as it solidifies, the arc is slowly brought back and another few drops of metal are deposited. Do not follow the up and down movement of the arc with the eyes; keep them on the molten metal.

8. When the arc is brought back to the now solidified puddle, it must be short, otherwise no metal will be desposited, the puddle will melt again, and it will drip.

9. It is important to realize that the entire process consists of slow, deliberate movements. There are no fast motions.

Vertical-down welding.   Vertical-down welding is applied at a fast pace. These welds are therefore shallow and narrow, and as such are excellent for sheet metal. This technique should not be used on heavy metal as the welds will not be strong enough. The general procedure follows:

1. Use ⅛- or ³⁄₃₂-in Fleetweld 180 electrode.

2. On thin metal, use 60 amps 16-gage sheets, 75 amps for 14-gage sheets.

3. Hold the electrode at a 30 to 45 degree angle as shown in Fig. 8.13 with the tip of the electrode pointing upward.

4. Hold a very short arc, but do not let the electrode touch the metal.

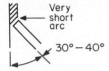

FIG. 8.13 Angle electrode is held in relationship to work for vertical-down welding. *(Lincoln Electric Co.)*

5. An up-and-down whipping motion will help prevent burn-through on very thin plate.
6. Watch the molten metal carefully.

The important consideration with vertical-down welding is to continue lowering the entire arm as the weld is made so the angle of the electrode does not change. Move the electrode so fast that the slag does not catch up with the arc. Since vertical-down welding gives thin, shallow welds, it should not be used on heavy material where large welds are required.

Overhead Welding

Various techniques are used for overhead welding. However, in general the following can be recommended as a beginning.

1. Use a ⅛-in (90 to 105 amps) or ³⁄₃₂-in (60 amps) electrode.
2. Put the electrode in the holder so it sticks straight out as shown in Fig. 8.14.

Side view

End view

FIG. 8.14 Electrode position for overhead welding. *(Lincoln Electric Co.)*

3. Hold the electrode at an angle approximately 30 degrees off vertical–both from the side and end of the work.

4. The major consideration with this type of welding is to hold a very short arc, since a long arc will cause the molten metal to fall.

5. If necessary, as dictated by the appearance of the molten metal, a slight back-and-forth motion along the seam with the electrode will help prevent dripping.

Welding Sheet Metal

Welding sheet metal presents an additional problem. The thinness of the metal makes it very easy to burn through. Follow these few simple rules:

1. Hold a very short arc.

2. Use ⅛- or ³⁄₃₂-in electrode.

3. Use low amperage (40 to 75), depending upon the electrode size.

4. Move fast. Don't keep the heat on any given spot too long.

5. Use lap welds whenever possible. This doubles the thickness of the metal.

Hard-Surfacing

There are several kinds of wear. The two most often encountered are metal-to-ground wear and metal-to-metal wear. Each of these types requires a different kind of hard-surfacing electrode. When applying the proper electrode, the service life of the part will more than double in most cases.

Follow this method for metal-to-ground hardsurfacing:

1. Grind the edge of the metal, until a 1-in-wide strip along the edge is bright in appearance.
2. Place the piece of metal on an incline of approximately 20 to 30 degrees. Choose the side of the work that will be subject to the most wear.
3. Use a ⅛-in Abrasoweld electrode, Fig. 8.15, at 75 to 90 amps. Strike the arc about 1 in from the edge of the metal.
4. The bead should be put on with a weaving motion, and it should be ½ to ¾ in wide. Do not let the arc blow over the edge, as this will dull the edge.
5. Use the back-stepping method of welding. Begin to weld 3 in from the heel of the share and weld to the heel. The second weld will begin 6 in from the heel, the third weld 9 in from the heel, etc.

Hard-surfacing items such as rollers for metal-to-metal wear is done in the following manner:

1. The part is inserted on a piece of pipe that is resting on two sawbucks. This enables the operator to turn the roller while welding.
2. Use Jet-LH BU-90 electrodes, ⁵⁄₃₂ in at 175 amps or ³⁄₁₆ in at 200 amps.
3. Weld across the wearing surface. Do not weld around.
4. Keep the roller cool by quenching with water and by stopping the welding periodically. This will prevent shrinking of the roller on the grease bearing.
5. Build up to the required dimension. The weld metal deposited

ELECTRODE IDENTIFICATION AND OPERATING DATA

The current ranges given represent minimum and maximum metered currents for which each electrode is designed. Actual machine setting is influenced by plate thickness, joint position and operator preference.

MILD STEEL

COATING COLOR	Conforms to Test Requirements of AWS Class	ELECTRODE BRAND NAME	ELECTRODE POLARITY (+) = "REVERSE" (−) = "STRAIGHT"	5/64" SIZE	3/32" SIZE	1/8" SIZE	5/32" SIZE	3/16" SIZE	7/32" SIZE	1/4" SIZE	5/16" SIZE
Light Tan	E6010	Fleetweld 5	DC (+)			75-130	90-175	140-225	200-275	250-325	280-400
Brick Red	E6010	Fleetweld 5P	DC (+)		40-75	80-135	110-180	155-250	225-295	245-325	
Tan	E6012	Fleetweld 7	DC (−) AC			90-150	120-200	170-275	250-335	250-335	
Light Tan	E6011	Fleetweld 35	AC DC (+)		75-105 70-95	75-120 70-110	90-145 80-145	120-180 110-150	150-235 135-180	190-300 170-270	
Red Brown	E6011	Fleetweld 35LS	AC DC (±)			80-130 70-120	100-150 90-135	120-160 100-145	150-200 135-180	185-260 185-235	
Dark Tan	E6013	Fleetweld 37	AC DC (±)		75-105 70-95	100-150 90-135	120-160 110-150	150-200 135-180	185-260 185-235	280-425 260-380	
Gray Brown	E7014	Fleetweld 47	AC DC (−)			110-160 100-145	150-225 135-200	200-280 180-250	260-340 225-305	280-425 260-380	360-460 330-430
Gray*	E6013	Fleetweld 57	AC DC (±)		75-105 70-95	100-150 90-135	150-200 135-180	200-260 180-240	250-310 225-280	300-360 270-330	
Brown	E6011	Fleetweld 180	AC DC (+)	45-80 40-75	40-90 40-80	60-120 50-110	115-150 105-135				
Gray	E7024	Jetweld* 1	AC DC (±)		65-120 60-110	115-175 100-160	190-240 160-215	240-300 220-280	300-380 270-340	350-440 320-400	
Brown	E6027	Jetweld 2	AC DC (±)			115-175 100-160	180-240 160-215	250-300 270-275	300-380 270-340	350-450 315-405	
Gray*	E7024	Jetweld 3	AC DC (±)			90-150 80-120	120-190 120-190	180-240 160-215	240-315 215-285	350-450 315-405	380-600 360-600
Gray	E7018	Jetweld LH-70	DC (+) AC		70-100 80-120	90-150 80-120	120-190 120-190	170-260 190-300	210-380 260-380	290-430 325-440	375-500 400-530
Gray	E7018	Jet LH® 72	DC (+) AC		70-100 80-120	70-100 80-120	130-190 135-225	170-260 180-280	250-330 270-370	300-400 325-420	
Gray	7018 (white numbers)	JET-LH 78	DC (+) AC			80-100	130-190 140-225	180-270 210-290	250-330 270-370	300-400 325-420	
Gray Brown	E7028	Jetweld LH-3800	AC DC (+)			80-100	180-270 170-240	240-330 210-300	275-410 260-380	360-520	

LOW ALLOY, HIGH TENSILE STEEL

Color	AWS Class	Lincoln Name	Current	1/8"	5/32"	3/16"	7/32"	1/4"
Pink	E7010-A1	Shield-Arc 85	DC(+)	75-130	90-175	140-225		
Pink*	E7010-A1	Shield-Arc 85P	DC(+)	75-130	90-185	140-225		250-350 / 250-400
Tan*	E7010-G	Shield-Arc HYP	DC(+)	75-130	90-185	140-225		250-350 / 300-400
Tan*	E7010-G	Shield-Arc 65+	DC(+)	75-130	90-185	140-225	160-250	
Gray	E8010-G	Shield-Arc 70+	DC(+)	75-130	90-185	140-225	160-250	
Gray-Brown	E8018-C1	Jet-LH 8018-C1	DC(+); AC	90-150 / 110-160	120-180 / 140-200	180-270 / 200-300		210-330 / 250-360
Gray-Brown	E8018-C3	Jet-LH 8018-C3	DC(+) AC	90-150 / 110-160	120-180 / 140-200	180-270 / 200-300		210-330 / 250-360
Gray	E8018-B2	Jetweld LH-90	DC(+) AC	110-160 / 130-170	120-200 / 140-230	160-260 / 200-300		190-310 / 240-350
Gray	E11018-M	Jetweld LH-110M	DC(+) AC	85-155 / 100-170	120-195 / 135-225	160-280 / 200-310		230-360 / 290-410

STAINLESS STEEL

Color	AWS Class	Lincoln Name	Current	5/64"	3/32"	1/8"	5/32"	3/16"	1/4"
Pale Green	E308-15	Stainweld 308-15	DC(+)	20-45	30-70	50-100	75-130	95-165	150-225
Gray	E308-16	Stainweld 308-16	DC(+); AC		30-60	55-95	80-135	115-185	200-275
Gray	E308L-16	Stainweld 308L-16	DC(+); AC		30-65	55-95	80-140	115-190	
Gray	E309-16	Stainweld 309-16	DC(+); AC		30-60	55-95	80-135	115-185	200-275
Pale Green	E310-15	Stainweld 310-15	DC(+)		30-70	45-95	80-135	100-165	
Gray	E310-16	Stainweld 310-16	DC(+); AC		30-65	55-100	80-140	120-185	200-275
Gray	E316-16	Stainweld 316-16	DC(+); AC		30-65	55-100	80-140	115-190	
Pale Green	E347-15	Stainweld 347-15	DC(+)		30-70	50-100	75-130	95-165	
Gray	E347-16	Stainweld 347-16	DC(+); AC		30-60	55-95	80-135	115-185	

BRONZE & ALUMINUM

Color	Class	Lincoln Name	Current	3/32"	1/8"	3/16"	1/4"
Peach	E CuSn-C	Aralweld●	DC(±)	20-55	50-125	70-170	90-220
White	Al-43	Aluminweld●	DC(+)		45-125	60-170	85-235

CAST IRON

Color	Class	Lincoln Name	Current	1/8" SIZE	5/32" SIZE	3/16" SIZE	1/4" SIZE
Light Tan	ESt	Ferroweld●	DC(±); AC	80-100	100-135 / 110-150	145-210 / 155-225	
Black	ENi-CI	Softweld●	DC(±) AC	60-110 / 65-120	120-180 / 135-230		

HARDSURFACING

Color	Lincoln Name	Current	1/8" SIZE	5/32" SIZE	3/16" SIZE	1/4" SIZE
Black	Abrasoweld●	DC(±)	75-250 / 80-165	110-250 / 120-275		150-375 / 165-410
Black	Faceweld● 1	DC(±); AC	40-50 / 50-165			
Black	Faceweld 12	DC(+); AC	60-150			
Dark Gray	Jet-LH BU-90	DC(±) AC	180-280 / 200-290		230-360 / 255-375	
Dark Gray	Mangjet●	DC(±) AC	160-260 / 185-285		200-350 / 200-385	
Dark Gray	Wearweld●	DC(+) AC	110-275 / 125-275		150-400 / 200-400	

IDENTIFICATION
Look for Lincoln's symbol of dependability.

IDENTIFICATION DOT
AWS Class (or NAMED) on each electrode. Exceptions: 3/32" & 1/8" Stainweld Faceweld 1 & 12 Aluminweld

Will operate on the AC-225-S.
DC only electrodes cannot be used on the AC-225-S.

●=Means registered Trademark of The Lincoln Electric Company
*=Has identification dot on coating

FIG. 8.15 Electrode identification and operating data. (*Lincoln Electric Co.*)

by BU-90 electrode is often so smooth that machining or grinding is not necessary.

During this welding process, the quenching of the workpiece also increases the hardness of the deposit.

Welding Cast Iron

When welding cast iron, the tremendous heat from the arc will be absorbed and distributed rapidly into the cold mass of the metal. This heating and sudden cooling creates white, brittle cast iron in the fusion zone. This is the reason that welds on cast iron often break. This fault can be overcome somewhat by following either of two methods:

1. Preheat the entire casting to 500 to 1200°F. If the cast iron is hot before welding, there will be no sudden chilling to create brittle white cast iron. The entire casting will cool slowly.
2. Weld only ½ in at a time, and do not weld at that same spot again until the weld is cool.

Unless a large furnace is available, the second method is more practical for machine-shop use. In either case, after welding cast iron, protect the casting against fast cooling. One method is to bury the piece in a sand or lime box. If sand or lime is not available, cover it with sheet metal or other nonflammable material that will exclude drafts and help to retain heat.

To prepare cast iron for welding, "vee" the joint out by grinding or filing to give complete penetration. This is especially important on thick castings where maximum strength is required. In some instances, a back-up strip may be used and plates may be gapped ⅛ in or more.

Cutting with Arc Welder

While most shops utilize an oxygen-acetylene torch for cutting, the arc welder in conjunction with an electrode can be used for cutting steel and cast iron by the following procedure:

1. Use ⅛- or ⁵⁄₃₂-in electrode.
2. Set welder on maximum amperage.
3. Hold long arc on edge of metal, melting it.
4. Push the arc through the molten metal, forcing it to fall away.
5. Raise the electrode, and start over again.

Continue this up and down sawing motion, melting the metal and pushing it away. If a lot of cutting becomes necessary, soak each electrode in water for a minute or two. This will keep it cooler, and the electrode will last longer.

Selecting Electrodes

The cost, quality, and appearance of the finished welding work depends a great deal on using the proper electrode. The chart in Fig. 8.16 gives electrode identification and operating data; Fig. 8.17 shows standard symbols for welding. A variety of welded joints are shown in Figs. 8.18 through 8.25.

Safety Precautions

To avoid possible injury while welding, the following precautions should be followed:

1. Provide protection from possible dangerous electrical shock.
   a. The electrode and work (or ground) circuits are electrically

Metal Thickness[1] (in.)	Weld Pass Number[1]	Electrode Dia. (in.)	Filler Rod Dia. (in.)	Tack Spacing (in.)	Alternating Current[2] (amp)	Arc Travel Speed (ipm)	Argon (cfh)	Shielding Gas Nozzle Dia. (in.)	Joint Details and Order of Welding	AWS Weld Symbol
3/64	1	1/16	3/32	3	55	11	10	3/8	0 MAX. ROOT OPENING	3/64
1/16	1	3/32	3/32	3	90	12.5	10	3/8		1/16
5/64	1	3/32	3/32	3	105	18	16	3/8	1/32 MAX. ROOT OPENING	5/64
3/32	1	1/8	1/8	3	120	12	16	3/8		3/32
1/8	1	1/8	1/8	4	155	10	16	3/8		1/8
3/16	1 B2R	3/16	3/16	4	230 230	12 12	16	1/2	ROOT OPENING 1/32 1/16 MAX.	1/8 BACK GOUGE 1/32

3/16	1 2	3/32	4	190 190	8 8	16	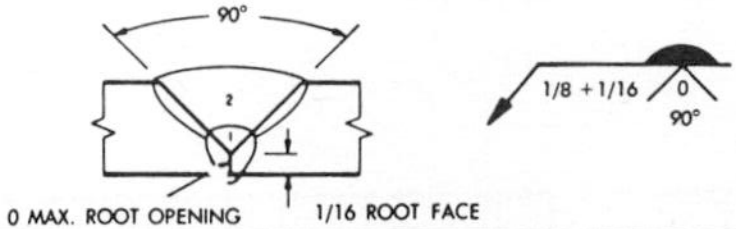 1/2
1/4	1 B2R	3/16	4	270 270	11 11	20	3/8 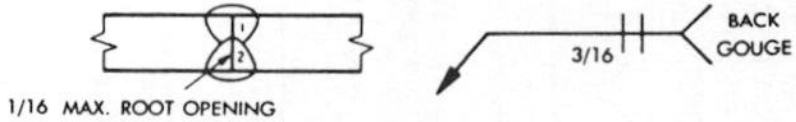
1/4	1 B2R	3/16	4	240 240	9 9	20	3/8 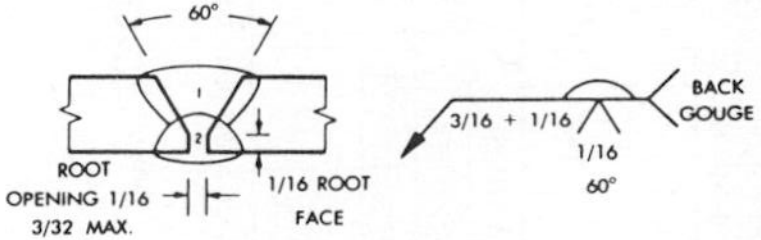
1/4	1 2	3/16	4	220 260	6 6	20	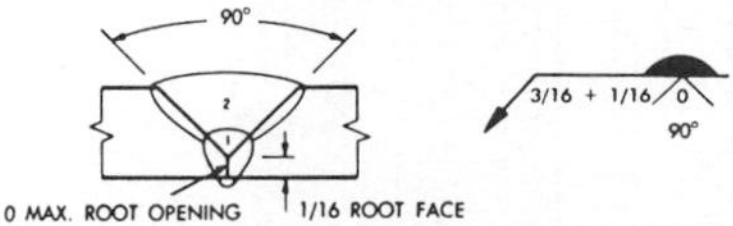 1/2

Diagram labels:

90° 1/16 ROOT FACE 0 MAX. ROOT OPENING — 1/8 + 1/16 / 0 / 90°

1/16 MAX. ROOT OPENING — BACK GOUGE 3/16

60° ROOT OPENING 1/16 3/32 MAX. 1/16 ROOT FACE — BACK GOUGE 3/16 + 1/16 / 1/16 / 60°

90° 0 MAX. ROOT OPENING 1/16 ROOT FACE — 3/16 + 1/16 / 0 / 90°

1 B indicates back gouge before making pass.
R indicates weld pass made on the reverse of the side on which the first pass was made.
2 Current should not vary more than 3% during welding.

FIG. 8.16 Typical tungsten–inert gas welding specifications.

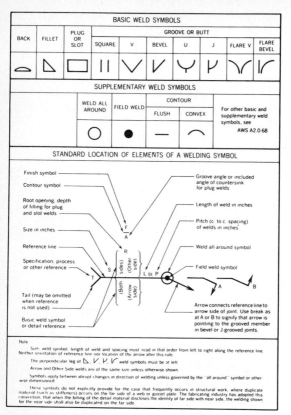

FIG. 8.17 Standard symbols for welding. [*R. O. Parmley (ed.), Field Engineer's Manual, McGraw-Hill, 1981.*]

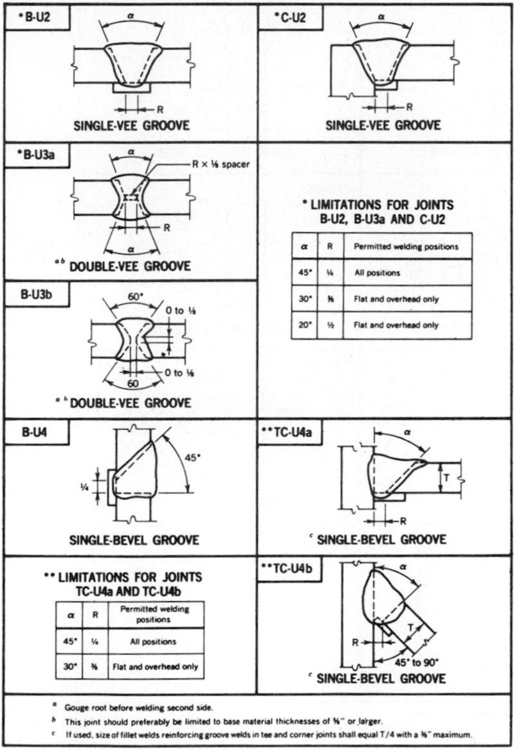
*B-U2
α
R
SINGLE-VEE GROOVE
*C-U2
α
R
SINGLE-VEE GROOVE
*B-U3a
α
R × ⅛ spacer
R
α
a b DOUBLE-VEE GROOVE
B-U3b
60°
0 to ⅛
0 to ⅛
60
a b DOUBLE-VEE GROOVE
B-U4
45°
¼
SINGLE-BEVEL GROOVE
**TC-U4a
α
T
R
c SINGLE-BEVEL GROOVE
**TC-U4b
α
T
R
45° to 90°
c SINGLE-BEVEL GROOVE

*** LIMITATIONS FOR JOINTS B-U2, B-U3a AND C-U2**

α	R	Permitted welding positions
45°	¼	All positions
30°	⅜	Flat and overhead only
20°	½	Flat and overhead only

**** LIMITATIONS FOR JOINTS TC-U4a AND TC-U4b**

α	R	Permitted welding positions
45°	¼	All positions
30°	⅜	Flat and overhead only

[a] Gouge root before welding second side.

[b] This joint should preferably be limited to base material thicknesses of ⅝" or larger.

[c] If used, size of fillet welds reinforcing groove welds in tee and corner joints shall equal T/4 with a ⅜" maximum.

FIG. 8.18 Manual shielded metal-arc welded joints of unlimited thickness. [*R. O. Parmley (ed.),* Field Engineer's Manual, *McGraw-Hill, 1981.*]

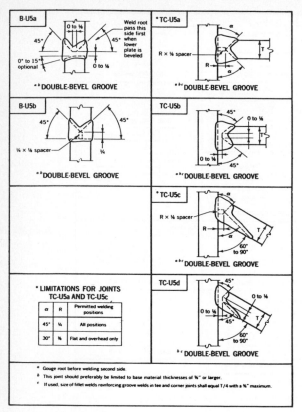

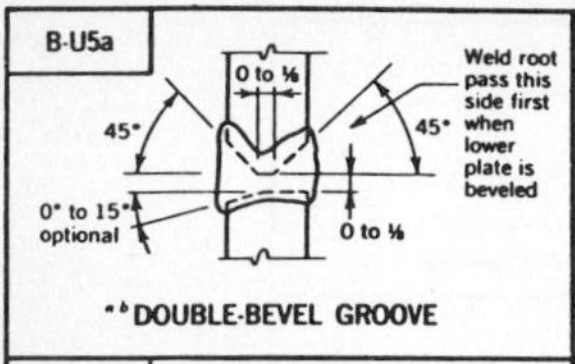
B-U5a
0 to 1/8
45°
45°
Weld root pass this side first when lower plate is beveled
0° to 15° optional
0 to 1/8
a b DOUBLE-BEVEL GROOVE

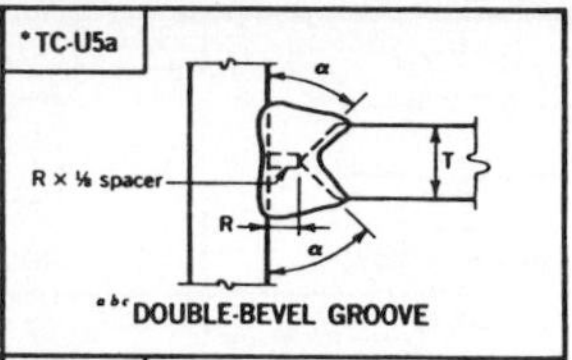
* TC-U5a
α
R × 1/8 spacer
T
R
α
a b c DOUBLE-BEVEL GROOVE

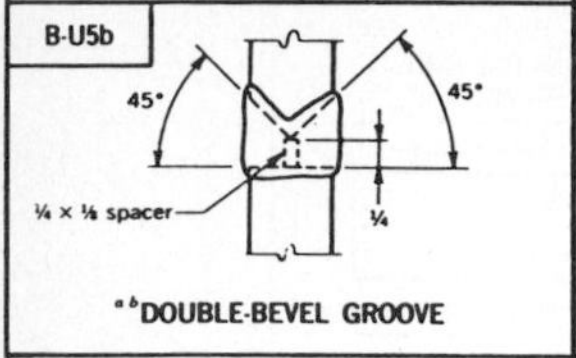
B-U5b
45°
45°
1/4 × 1/8 spacer
1/4
a b DOUBLE-BEVEL GROOVE

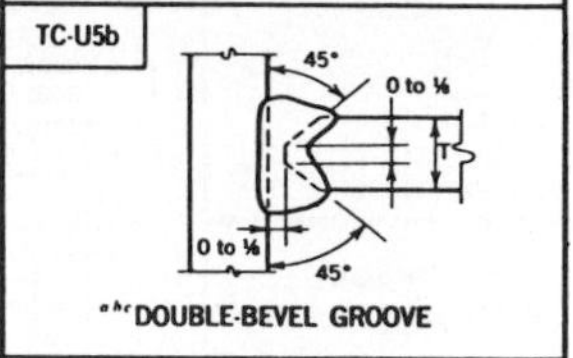
TC-U5b
45°
0 to 1/8
T
0 to 1/8
45°
a b c DOUBLE-BEVEL GROOVE

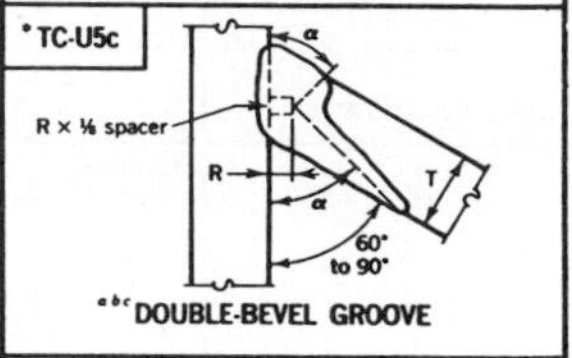
* TC-U5c
α
R × 1/8 spacer
R
T
α
60° to 90°
a b c DOUBLE-BEVEL GROOVE

*** LIMITATIONS FOR JOINTS TC-U5a AND TC-U5c**

α	R	Permitted welding positions
45°	1/4	All positions
30°	3/8	Flat and overhead only

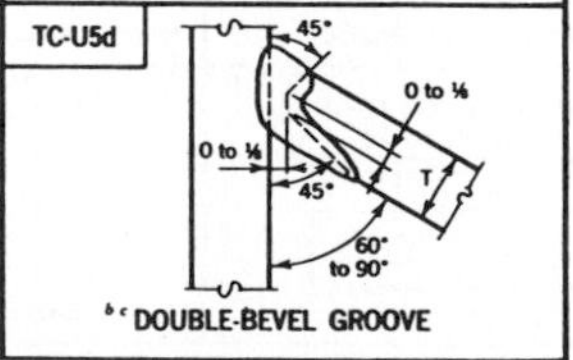
TC-U5d
45°
0 to 1/8
0 to 1/8
45°
T
60° to 90°
b c DOUBLE-BEVEL GROOVE

[a] Gouge root before welding second side.

[b] This joint should preferably be limited to base material thicknesses of 5/8" or larger.

[c] If used, size of fillet welds reinforcing groove welds in tee and corner joints shall equal T/4 with a 3/8" maximum.

FIG. 8.18 *(Continued).*

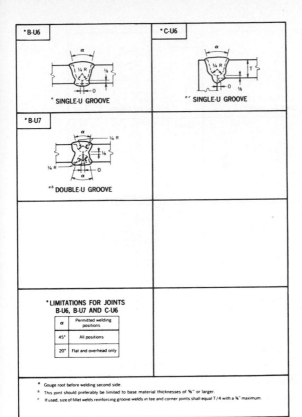

*B-U6	*C-U6
α ¼ R ⅛ O ᵃ SINGLE-U GROOVE	α ¼ R T O ⅛ ᵃᶜ SINGLE-U GROOVE
*B-U7 α ¼ R ¼ R ⅛ α O ᵃᵇ DOUBLE-U GROOVE	
*LIMITATIONS FOR JOINTS B-U6, B-U7 AND C-U6 | α	Permitted welding positions | |---|---| | 45° | All positions | | 20° | Flat and overhead only |	

ᵃ Gouge root before welding second side.

ᵇ This joint should preferably be limited to base material thicknesses of ⅜" or larger.

ᶜ If used, size of fillet welds reinforcing groove welds in tee and corner joints shall equal T/4 with a ⅜" maximum.

FIG. 8.18 *(Continued).*

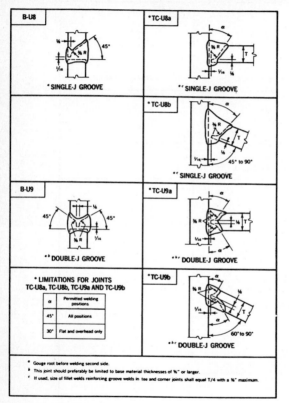

B-U8

* SINGLE-J GROOVE

*TC-U8a

** SINGLE-J GROOVE

*TC-U8b

** SINGLE-J GROOVE

B-U9

** DOUBLE-J GROOVE

*TC-U9a

***ᵇᶜ DOUBLE-J GROOVE

*** LIMITATIONS FOR JOINTS
TC-U8a, TC-U8b, TC-U9a AND TC-U9b**

α	Permitted welding positions
45°	All positions
30°	Flat and overhead only

*TC-U9b

***ᵇᶜ DOUBLE-J GROOVE

ᵃ Gouge root before welding second side.

ᵇ This joint should preferably be limited to base material thicknesses of ⅜″ or larger.

ᶜ If used, size of fillet welds reinforcing groove welds in tee and corner joints shall equal T/4 with a ⅜″ maximum.

FIG. 8.18 *(Continued).*

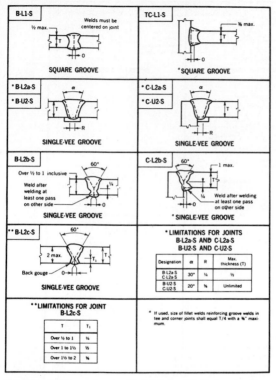

FIG. 8.19 Submerged arc welded joints of limited and unlimited thickness. [*R. O. Parmley (ed.), Field Engineer's Manual, McGraw-Hill, 1981.*]

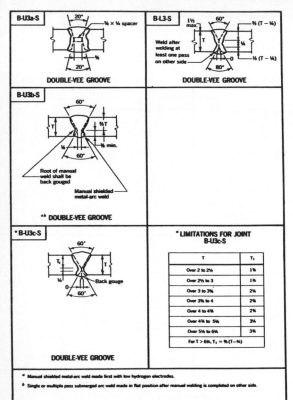

B-U3a-S

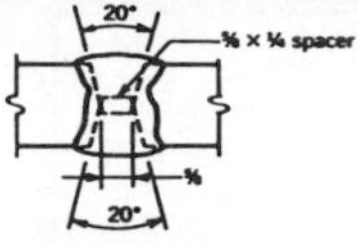
20°
5/8 × 1/4 spacer
5/8
20°

DOUBLE-VEE GROOVE

B-L3-S

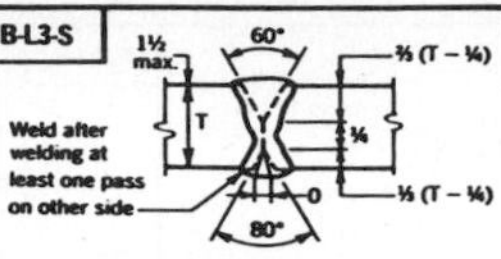
1 1/2 max.
60°
2/3 (T − 1/4)
T
1/4
Weld after welding at least one pass on other side
0
1/3 (T − 1/4)
80°

DOUBLE-VEE GROOVE

B-U3b-S

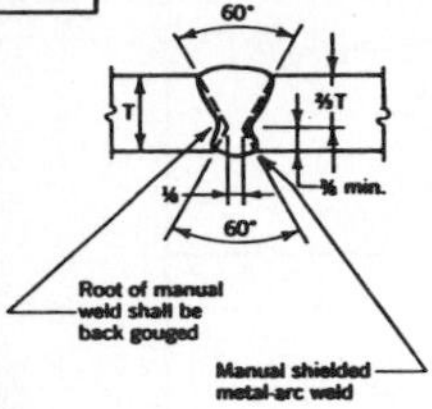
60°
T
2/3 T
1/4
5/8 min.
60°
Root of manual weld shall be back gouged
Manual shielded metal-arc weld

[a b] DOUBLE-VEE GROOVE

***B-U3c-S**

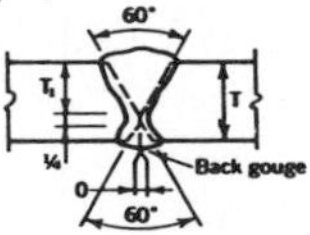
60°
T_1
T
1/4
Back gouge
0
60°

DOUBLE-VEE GROOVE

***LIMITATIONS FOR JOINT B-U3c-S**

T	T_1
Over 2 to 2 1/2	1 3/8
Over 2 1/2 to 3	1 3/4
Over 3 to 3 5/8	2 1/8
Over 3 5/8 to 4	2 3/8
Over 4 to 4 3/4	2 3/4
Over 4 3/4 to 5 1/2	3 1/4
Over 5 1/2 to 6 1/4	3 3/4
For $T > 6\frac{1}{4}$, $T_1 = \frac{2}{3}(T - \frac{1}{4})$	

[a] Manual shielded metal-arc weld made first with low hydrogen electrodes.

[b] Single or multiple pass submerged arc weld made in flat position after manual welding is completed on other side.

FIG. 8.19 *(Continued).*

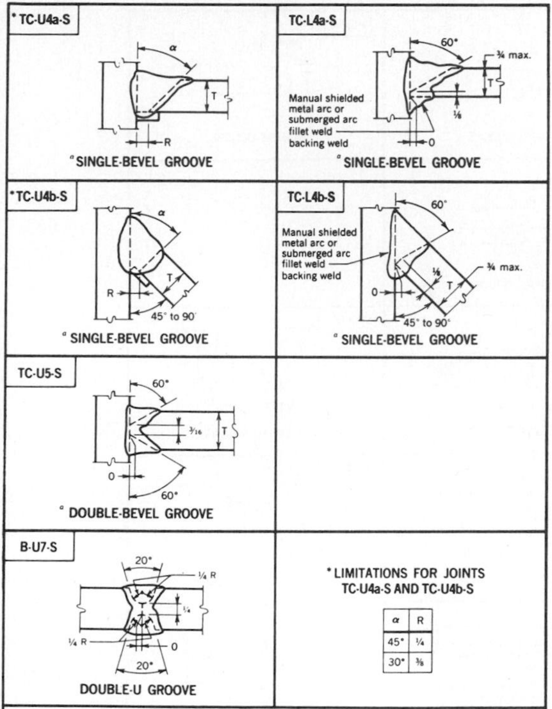
*TC-U4a-S
α
T
R
[a] SINGLE-BEVEL GROOVE
TC-L4a-S
60°
¾ max.
T
⅛
Manual shielded metal arc or submerged arc fillet weld backing weld
0
[a] SINGLE-BEVEL GROOVE
*TC-U4b-S
α
T
R
45° to 90°
[a] SINGLE-BEVEL GROOVE
TC-L4b-S
60°
Manual shielded metal arc or submerged arc fillet weld backing weld
⅛
¾ max.
T
0
45° to 90°
[a] SINGLE-BEVEL GROOVE
TC-U5-S
60°
3/16
T
0
60°
[a] DOUBLE-BEVEL GROOVE
B-U7-S
20°
¼ R
¼
¼ R
0
20°
DOUBLE-U GROOVE

*LIMITATIONS FOR JOINTS TC-U4a-S AND TC-U4b-S

α	R
45°	¼
30°	⅜

[a] If used, size of fillet welds reinforcing groove welds in tee and corner joints shall equal T/4 with a ⅜" maximum.

FIG. 8.19 *(Continued).*

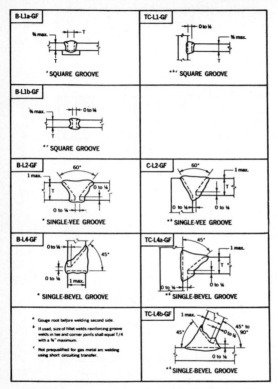

FIG. 8.20 Gas metal and flux cored arc welded joints of limited thickness. [*R. O. Parmley (ed.), Field Engineer's Manual, McGraw-Hill, 1981.*]

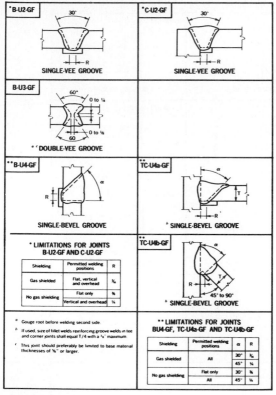

FIG. 8.21 Gas metal and flux cored arc welded joints of unlimited thickness. [*R. O. Parmley (ed.),* Field Engineer's Manual, *McGraw-Hill, 1981.*]

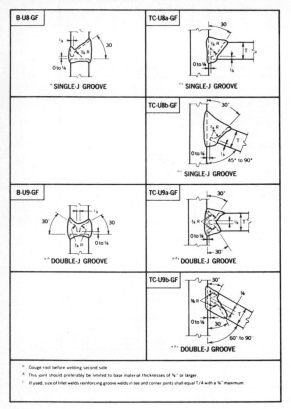

FIG. 8.21 *(Continued).*

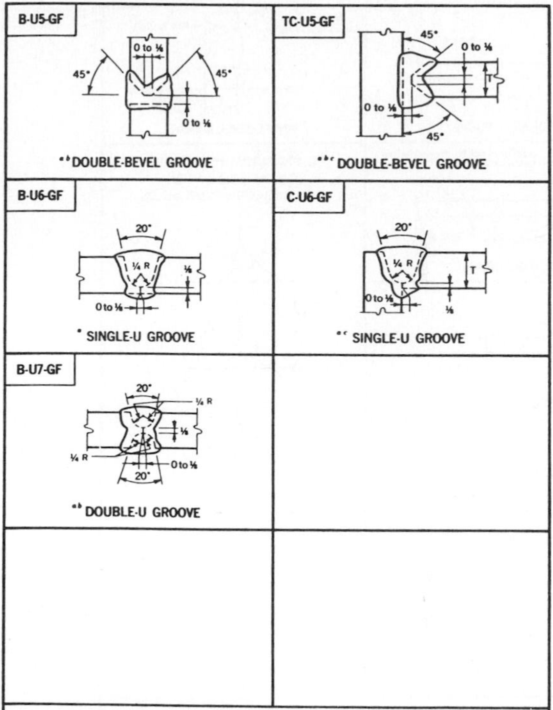
B-U5-GF
0 to 1/8
45°
45°
0 to 1/8
a b DOUBLE-BEVEL GROOVE
TC-U5-GF
45°
0 to 1/8
T
0 to 1/8
45°
a b c DOUBLE-BEVEL GROOVE
B-U6-GF
20°
1/4 R
1/8
0 to 1/8
a SINGLE-U GROOVE
C-U6-GF
20°
1/4 R
T
0 to 1/8
1/8
a c SINGLE-U GROOVE
B-U7-GF
20°
1/4 R
1/8
1/4 R
0 to 1/8
20°
a b DOUBLE-U GROOVE

[a] Gouge root before welding second side.

[b] This joint should preferably be limited to base material thicknesses of 5/8" or larger.

[c] If used, size of fillet welds reinforcing groove welds in tee and corner joints shall equal T/4 with a 3/8" maximum.

FIG. 8.21 *(Continued).*

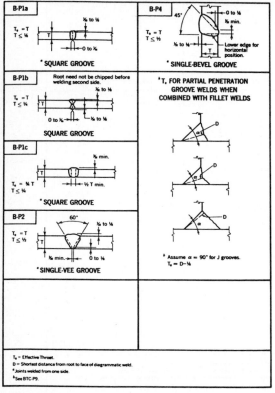

FIG. 8.22 Manual shielded metal-arc welded joints. [*R. O. Parmley (ed.), Field Engineer's Manual, McGraw-Hill, 1981.*]

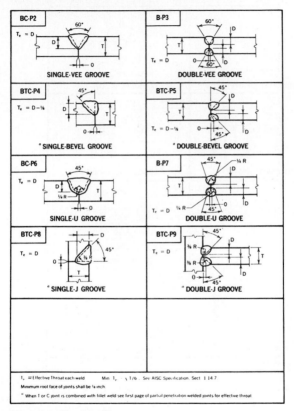

BC-P2	B-P3
$T_e = D$ 60° SINGLE-VEE GROOVE	60° $T_e = D$ 60° DOUBLE-VEE GROOVE
BTC-P4 $T_e = D - \frac{1}{8}$ 45° " SINGLE-BEVEL GROOVE	BTC-P5 45° $T_e = D - \frac{1}{8}$ 45° " DOUBLE-BEVEL GROOVE
BC-P6 $T_e = D$ 45° ¼ R SINGLE-U GROOVE	B-P7 45° ¼ R $T_e = D$ ¼ R 45° DOUBLE-U GROOVE
BTC-P8 $T_e = D$ ⅜ R 45° " SINGLE-J GROOVE	BTC-P9 45° $T_e = D$ ⅜ R ⅜ R 45° " DOUBLE-J GROOVE

T_e = Effective Throat each weld   Min T_e = 1/6 . See AISC Specification. Sect 1.14.7

Minimum root face of joints shall be ⅛ inch.

" When T or C joint is combined with fillet weld see first page of partial penetration welded joints for effective throat

FIG. 8.22 *(Continued).*

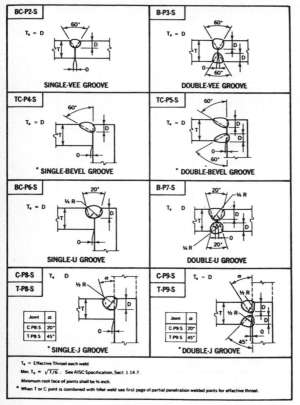

FIG. 8.23 Submerged arc welded joints. [*R. O. Parmley (ed.),* Field Engineer's Manual, *McGraw-Hill, 1981.*]

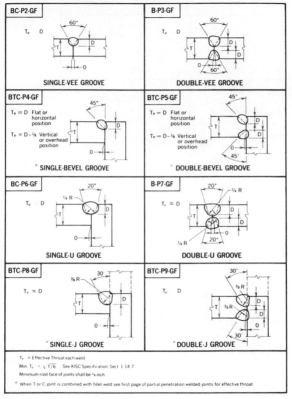

FIG. 8.24 Gas metal and flux cored arc welded joints. [*R. O. Parmley (ed.), Field Engineer's Manual, McGraw-Hill, 1981.*]

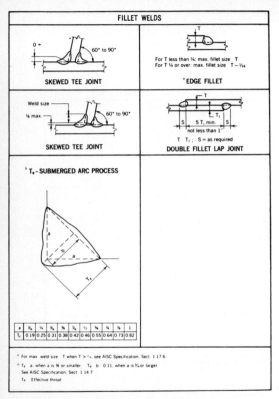

FIG. 8.25 Fillet welds for manual shielded metal-arc, submerged arc, gas metal, and flux cored welded joints. [*R. O. Parmley (ed.),* Field Engineer's Manual, *McGraw-Hill, 1981.*]

"hot" when the welder is on. Never permit contact between "hot" parts of the circuits and bare skin or wet clothing. Wear dry, hole-free gloves to insulate hands.

b. Always insulate the operator from work and ground using dry insulation when welding in damp locations, on metal decks or floors, on gratings or scaffolds, and particularly in places where the operator's body can possibly be in contact with grounds.

c. Maintain the electrode holder, work clamp, welding cable, and welding machine in good, safe operating condition.

d. Never dip the electrode holder in water for cooling.

e. Never simultaneously touch electrically "hot" parts of electrode holders connected to two welders because voltage between the two can be the total of the open circuit voltage of both welders.

2. When working above floor level, protect the operator from a fall should an electrical shock occur. Never wrap the electrode cable around any part of the body.

3. Arc burn may be more severe than sunburn. Therefore:

a. Use a shield with proper filter and cover plates to protect eyes from sparks and the rays of the arc when welding or observing open arc welding. Filter lens should be 10, 11, or 12 and conform to ANSI Z87.1 standards.

b. Use suitable clothing to protect the welder's skin and that of helpers from the arc rays.

c. Protect nearby personnel with suitable nonflammable screening and/or warn them not to watch the arc or expose themselves to the arc rays or to hot spatter of metal.

4. Droplets of molten slag and metal are thrown or fall from the welding arc. Protect with oil-free protective garments such as leather gloves, heavy shirt, cuffless trousers, high shoes, and a

cap over the hair. Wear ear plugs when welding out of position or in confined places.

5. Always wear safety glasses in a welding area. Use glasses with side shields near slag chipping operations.

6. Remove fire hazards well away from the area. If this is not possible cover them to prevent the welding sparks from starting a fire. The welding sparks and hot material from welding can easily go through small cracks and openings to adjacent areas.

7. When not welding, place the holder where it is insulated from the work and ground system. Accidental grounding can cause overheating and create a fire hazard.

8. Be sure the work cable is connected to the work as close to the welding area as practical. Work cables connected to the building framework or other locations some distance from the welding area increase the possibility of the welding current passing through lifting chains, crane cables, or other alternating circuits. This can create fire hazards or overheat lifting chains or cables until they fail.

9. Welding may produce fumes and gases hazardous to health. Avoid breathing these fumes and gases. When welding, keep head out of the fumes. Use enough ventilation and/or exhaust at the arc to keep fumes and gases from the breathing zone. When welding on galvanized, lead-plated, or cadmium-plated steel and other metals which produce toxic fumes, even greater care must be taken.

10. Do not weld in locations near chlorinated hydrocarbon vapors coming from degreasing, cleaning, or spraying operations. The heat and rays of the arc can react with solvent vapors to form phosgene—a highly toxic gas—and other irritating products.

11. Do not heat, cut, or weld tanks, drums, or containers until the proper steps have been taken to ensure that such procedures will not cause flammable or toxic vapors from substances inside. They can cause an explosion even though they have been "cleaned."

12. Vent hollow castings or containers before heating, cutting, or welding. They may explode.

13. For more detailed information, obtain a copy of "Safety in Welding and Cutting," ANSI Standard Z49.1, from the American Welding Society, Miami, FL 33125.

Other welding methods, such as forge welding, cold welding, gas welding, tungsten–inert gas welding, metal–inert gas welding, and spot welding are too lengthy to cover in a book of this type. Detailed information may be found in books on the respective subjects.

9

Heat Treating Steel

Most tools currently is use in any machine shop must be properly heat-treated to withstand the job they are to do. These tools include mills, reamers, twist drills, punches—even the simple screwdriver. All must undergo some form of treatment before being put to use. Such treatment can include hardening, tempering, annealing, forging, normalizing, and carburizing. In general, the heat treatment is applied to carbon and alloy steels mostly, but other steels are also used and subjected to one form of treatment or other.

HARDENING

Hardening of steel is the process of imparting certain physical properties to the steel so it will resist wear, have tensile strength so it will not stretch or strain, and have sufficient impact strength to stand blows. The degree of heat and the length of time the heat is applied in each case depend on the type of material, the shape and size of the workpiece, and the purpose for which it is to be used.

Hardening of steel consists mainly of heating the part to a predetermined temperature for a predetermined length of time. The temperature at which steel will harden most efficiently is called the critical temperature or point of recalescence. In some cases, the part may be heated somewhat beyond this temperature, then quickly immersed in water or brine, which is known as the quench. However, many tool steels are quenched in oil, or a combination of oil and water, while others are allowed to cool in still air. This latter method is especially true of certain springs.

TEMPERING

All steel tools that have been heated to high temperatures and immediately quenched have developed internal strains. When exposed to sudden changes in temperature, the steel is likely to crack unless these strains are relieved by the heat-treating procedure known as tempering—sometimes called "drawing." The procedure consists of reheating the steel to a comparatively lower temperature than it was subjected to during the hardening process, usually between 300 and 600°F, depending on the degree of hardness desired. It is also common practice to draw the metal beyond this point in certain cases. Once the tempering temperature has been reached and maintained for the predetermined time, the part should not be quenched as before; rather, it should be allowed to cool in the air.

ANNEALING

Hardened steel is normally not in a machinable state and must therefore be annealed during the work stage, and then rehardened once the machining operation has been completed. Annealing is done by applying heat at a temperature of about 15 to 100°F below the critical point. The process, when performed correctly, demands a slow and increasing heat until the desired temperature has been reached. Then the part is allowed to cool slowly, never quenched.

FORGING

In most cases, steel cannot be shaped without first being heated to forging temperatures. Parts to be forged should be heated slowly to the required temperature, which is usually from 200 to 350°F above the critical point. After removing the steel from the furnace or forge, it is allowed to cool in still air with all drafts and dampness being avoided.

NORMALIZING

This is a process quite similar to tempering, but the term is usually applied to work that has been forged. Most modern steels do not require this process unless excessive forging strains have been applied. If so, normalizing requires a temperature of from 50 to 75°F above the critical point. Gun barrels, for example, that have been forged, bored, and rifled often are normalized to relieve stresses so they will maintain straightness when tapered.

CARBURIZING

Carburizing is a heat-treating process that uses certain special gases to introduce an extra amount of carbon to low-grade steels so they can be used as substitutes for more expensive steels. This process is also known as case hardening because it is a surface-hardening process. The resulting carbon content and depth of hardening or "case" will depend on the gas mixture, type of equipment, and heating time.

Work to be case-hardened must be machined to dimensions, yet allowances for ground finishes can be made by estimating and controlling the depth of case, which may be $\frac{3}{16}$ in and over. If the work is executed properly, the result is a uniform glass-hard exterior section with no effects on the core or internal structure of the steel. A quench bath is normally used after each part or batch of parts has been heated for the required period of time.

Carburizing is not the only form of surface hardening. Nitriding is a similar process using ammonia gas as a treating agent. There are also many other methods in current use. The use of a cyanide bath is frequently made when a shallow case is desired, but this chemical is extremely dangerous to use and only experienced persons should even consider its use.

Flame-hardening or coloring is another method used to harden steel, especially where resistance to surface wear is the main objective. Induction hardening is another development which is accomplished by placing the work inside a magnetic field set up by an inductor coil which carries high-frequency currents. The heating and quenching cycle takes only a few seconds for each part.

A cold-treating furnace is often used to stabilize the steel used in precision tools and instruments. The "furnace" is actually a carbon-dioxide refrigerator in which the temperature can be reduced to $-120°F$. This treatment stabilizes the steel and offsets the effects of subsequent temperature changes.

HEAT FOR HEAT TREATING

There are, and have been, several methods of applying heat to steel for the purpose of treating it. Open fires, forges in the old blacksmith's shops, and the like were quite common in the last centruy. In recent times, however, torches or heat-treating furnaces are the most common.

There are at least two distinct types of heat-treating furnaces: the "box type" and the pot or crucible type. Both can be used for all the heat-treating processes previously mentioned. Sources of heat can be natural gas, LP gas, charcoal, electricity, or oil. Of these fuels, electricity, in electric furnaces, is the most popular for small parts.

With the box-type furnace, the heat treating is done directly by the heat that is produced within the chamber. With the crucible type, the part or parts to be treated are immersed in a red-hot lead or salt bath. The size of both will vary from huge devices suitable for heat treating cannon barrels to small bench-top models used for delicate parts for instruments and firearms.

THERMOCOUPLE PYROMETERS

Although many shops build simple heat-treating furnaces for certain projects, most are equipped with some means of measuring the temperature. The most common means is a thermocouple pyrometer with an automatic heat control to maintain the desired temperature for the entire heating period. In general, thermocouple pyrometers consist of two lengths or wire of different metals or alloys, separated from each other except at one end, which is securely twisted and welded so both metals will unite. The free ends of the wires are connected to a dc millivoltmeter while the welded end is inserted where the temperature is to be measured. The heat at this end causes a difference in electric potential, which is indicated on the millivoltmeter, the scale of which is calibrated in degrees.

HARDNESS TESTER

The degree of hardness of a heat-treated part is measured by several methods, but the Rockwell hardness tester is the most commonly used instrument in the United States. The degree of hardness is calibrated on different scales, such as A, B, C, etc., and are read, for example, as 70 on the Rockwell C scale, 95 on the Rockwell B scale, etc.

The Rockewll hardness tester is a precision machine which measures on a graduated dial scale the linear increment of depth penetration made by a standard penetrator acting under a minor or initial load of 10 kg and a major load of either 60, 100 or 150 kg. There are also other types of testers used in hardness testing.

For testing soft steel the penetrator of this instrument is normally a $\frac{1}{16}$-in ball, and the major load is 100 kg. For testing hard materials and hardened tools, a 120 degree diamond cone is used with a major load of 150 kg. The minor load remains the same in either case.

The operation of the Rockwell hardness tester is as follows:

1. Select the proper penetrating point, clean it thoroughly, and place in the head plunger rod and tighten in place.

2. Place proper anvil on the elevating screw after it has been thoroughly cleaned.

3. Place work to be tested securely upon anvil or table.

4. Elevate specimen into contact with penetrator and turn the elevating screw until the small pointer is nearly vertical, and slightly to the left of the dot. Then turn the screw slightly more until the large pointer is pointing upward.

5. Turn the bezel of the dial gage until the Set arrow on scale is exactly behind the larger pointer.

6. Push crank handle backward past dead center until the pointer on the dial comes to rest. This procedure has applied the load of 100 kg.

7. Pull crank handle to the starting position. This lifts the major load, but leaves the minor load on.

8. Read Rockwell C hardness number, lower the elevating screw, and remove the work from the machine.

Another type of tester is the scleroscope hardness tester. Hardness values on this tester are obtained by observing the height of rebound of a diamond-tipped hammer which is allowed to fall through a fixed distance upon the polished tool surface. The hardness is read from a dial positioned on the instrument.

To operate, lower the instrument on the work held in the anvil. Turn the knurled knob clockwise for about five-eighths of a turn or until interval stop is reached. At this point the rebound occurs, which is indicated on the dial. Releasing the instrument returns it for another cycle of operation. The instrument has the capacity for about 1000 tests per hour. The hardness values are indicated on a dial scale graduated from 0 to 140.

There is a relation between Rockwell C and scleroscope hardness. For 62 Rockwell C, the scleroscope reading is 86; for 65 Rockwell C, scleroscope reading is 92. Table 9.1 gives the relative values for Rockwell C hardness, and scleroscope hardness numbers for some values applicable to hardened tools.

TABLE 9.1

Rockwell C hardness	Scleroscope hardness
50	68
51	71
53	73
55	75
56	78
58	81
60	84
62	86
63	89
65	92

10

Metal Finishes

Most metals, other than stainless steels, are given some type of finish before they are put into use. Metal finishes offer protection to the metal and also give it a pleasing appearance. In practice, the finish will vary from a simple priming and painting to jeweling or polishing to blueing or blackening by oxidation. Some metals are plated with chrome, nickel, or other substance for extreme protection. The exact method used will depend on the type of metal and for what purpose it will be used.

DAMASCENING METAL

This process involves overlapping spot-polish marks on metal that give a hammered effect. The process is also called "jeweling" and "engine turning."

For small jobs, an engine turning chuck is normally used. This tool consists of a steel holder with about a ⅛-in shank and an abrasive charged rubber tip. The specially made abrasive tip (Fig. 10.1) gives an even impression on steel and simplifies the work of engine turning and eliminates the danger of cutting deep rings.

Many shops use engine turning brushes rather than abrasive tips. These brush wires follow the contours of rounded parts better, giving a complete jeweled pattern. They are also reported to be superior on flat surfaces with long tool life. Silicon carbide abrasive and oil are used in conjunction with the wire brush to obtain the pattern. Drill press rpm should be relatively high—above 2000 rpm.

When making engine turning designs on parts that require precise patterns, specially designed fixtures are normally used to obtain a symmetrically jeweled pattern. These fixtures utilize an indexed scale to ensure that the swirled spots will be symmetric and produce the desired pattern.

To use, secure the object to be jeweled in the fixture, and posi-

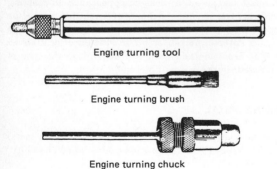

Engine turning tool

Engine turning brush

Engine turning chuck

FIG. 10.1 Abrasive tips used for engine turning.

tion it under the spindle of the drill press. Coat the metal with an abrasive compound and start the pattern, by bringing the engine turning tip or wire brush down on the metal to make the first spot. Then adjust the fixture for the position of the next spot, which is usually a distance equal to one-half the diameter of the first spot. Bring the brush down again on the metal to overlap the first spot. Continue this procedure until one row of spots is completed. Then the metal object is moved or rotated approximately one-half spot diameter and the process repeated for another row. This continues until the entire object has been fully jeweled in the areas desired.

BLUEING

A metal finish known in the United States as "blueing" has been in use for the past century or so mainly to color metal used in firearms and hand tools. The finished product, however, is seldom a true blue; rather, the color will be either blue-black or a deep black. The term blueing probably originated from the temper blue often given some metal parts from time to time. The British still call this method "browning" although the resulting color is really blue-black to black. Prior to the nineteenth century, most coloring given to metals was a brown to brownish black in color, so the term browning probably remains for this reason.

In general, there are four basic methods of blueing steel:

1. Instant cold blue
2. Slow-rust method
3. Hot-water method
4. Hot-caustic method

Of these four, the hot-caustic method is currently used most often in all metal-finishing areas. One such finish is called Pentrate and is available from Heatbath Corporation of Springfield, MA.

The Pentrate process produces a protective penetrating finish on iron and steel parts including normal SAD and NE steels, highly alloyed nonstainless steels, high-speed steels, cast iron, and wrought iron. The surface condition of an article determines the exact final finish—that is, whether it has a matt, glossy, or somewhere-in-between finish. In all cases, the work to be treated must be absolutely clean and free from oil, grease, rust, and scale.

The Pentrate process will have no effect on any previous heat treatment. When properly applied, the Pentrate process produces a penetration of .00035 to .00045 in and the resulting deep black finish will not chip, crack, peel, flake, or rub off under normal usage.

This process, however, will not work on nonferrous metals such as lead, aluminum, brass, copper, tin, and zinc, and such metals can seriously contaminate the baths. Even baskets or fixtures for holding work must be iron or steel, and joints should be welded, never soldered.

The Pentrate process is carried out in tanks such as those shown in Fig. 10.2. There are usually six tanks, and work progresses from left to right:

1. Hot alkali cleaner

2. Cold running water, rinse

3. Blueing solution (lower temperature)

4. Blueing solution (higher temperature)

5. Cold running water, rinse

6. Hot soluble oil bath

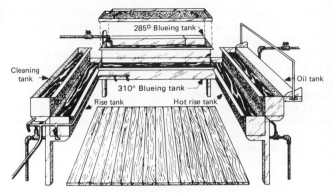

FIG. 10.2 Hot caustic blueing apparatus.

The first tank contains the cleaner, and any good commercial alkali cleaner will suffice. Usually a mixture of 6 to 8 oz of cleaner per gallon of water is sufficient to remove dirt, oil, and grease. Drain and refill this tank frequently. Work cannot be thoroughly cleaned if the surface of the solutions shows a quantity of oil and dirt.

The second tank is supplied with water inlet and a baffle-type overflow. When in operation, water should be allowed to flow in and out freely.

Tank no. 3 contains the blueing solution. The final working level of the blueing solution should be approximately 6-in below the top of the tank to allow ample room for displacement when the work is immersed. On this basis, each tank is filled one-quarter full of water and heated slowly. The blueing salts are added in small quantities, allowing each quantity to dissolve before making further additions. Continue to add salts and water until the solution is visibly boiling at 285°F. This will require approximately 6.5 lb of blueing salts for each gallon of water used.

The fourth tank also contains the blueing solution, and the procedure is exactly the same as for the third tank except that a higher temperature is desired—approximately 310°F. This additional temperature is reached by adding more blueing salts.

The correct temperatures of both tanks 3 and 4 are maintained thereafter by careful replacement of water lost through evaporation. This water replacement must be made slowly and only through splash guards provided. The salts are added only to compensate for loss due to mechanical dragout and to maintain the desired working level. In other words, the temperatures of these two tanks are not controlled by heat; rather by the chemicals and water.

Tank no. 5 is similar to tank no. 2. Continuous running water is essential.

The final tank (no. 6) can be filled with water to a point just below the desired level, then heated slowly. Add soluble oil to a consistency desired for the work at hand. Usually 1 part soluble oil and 1 part water is adequate. If a very oily finished part is desired, add more oil. If a comparatively dry surface is desired, add more water. The solution in this tank need not boil too vigorously. A slow "rolling boil" is sufficient. In heating oil, however, make absolutely certain that the oil is designed for this type of work. Many fires have been started by applying heat to oil and extreme precautions must be taken.

After the work has been removed from tank no. 6, the hot water will evaporate, leaving a slightly oiled surface. If an even more rust-resistant finish is desired, use Heatbath Pen-Dip oil no. 300.

If parts are to be painted, the soluble oil may be washed off and rinsed in tanks 1 and 2 and then dried thoroughly in sawdust or by heating.

All tanks now being at the desired working level, actual production should proceed as follows:

1. The majority of steel parts have the benefit of the full blueing treatment after a 15-minute immersion in each of the two tanks. It is good practice, therefore, to let the immersion time in each of the first two tanks govern the length of immersion time in each of the other tanks.

2. Place work to be treated in steel wire mesh baskets using as open a mesh as possible.

3. Immerse work in tank no. 1. Have cleaning solution boiling vigorously. Allow work to remain 15 minutes. Next rinse thoroughly in tank no. 2.

4. Remove work from tank no. 2 and place in tank no. 3. In doing so, the parts should be lowered slowly, as cold water coming in contact with boiling blueing solution will spatter. Be sure work is completely submerged in the blueing solution.

5. After 15 minutes transfer work from tank no. 3 directly to tank no. 4 and allow to remain 15 minutes.

6. Work is now removed from tank no. 4 and placed in tank no 5 (cold rinse) as rapidly as possible. Be sure work is thoroughly rinsed in this tank. Work in baskets should be shaken to assure complete washing. If parts being treated are assemblies or have blind holes or complicated grooves, a boiling water rinse is required.

7. After thorough rinsing, work is placed in tank no. 6. The soluble oil solution should be boiling but not vigorously.

8. When work is removed from tank no. 6, it should be allowed to dry.

In performing the above operation, do not attempt to plug up holes with rubber, asbestos, or other materials. This practice is unnecessary and dangerous, as steam pockets may develop, blowing the plug out and causing violent splashing of the solution.

The hot-caustic blueing method is not dangerous if certain precautions are taken. The greatest danger is the hot lye solution, which can cause burns and blindness. Always wear goggles or a suitable shield. Rubber gloves are obviously required when working around any hot alkali solution. A rubber apron will protect clothing and prevent serious burns. Rubber shoes are also recommended, as the blueing solution will attack leather as well as clothing.

In the event of burns from the blueing solution, wash thoroughly in cold water and apply boric acid (either in powder or liquid form) or any approved salve. If the burn is serious, seek a doctor's aid.

PARKERIZING

Parkerizing has long been the metal finishing process used by the armed forces since it is highly rust resistant and wears longer than most hot-caustic blueing processes. The old method consisted of boiling the parts to be finished in a solution of Parko powder composed of specially prepared powdered iron and phosphoric acid. During the process, minute particles of the metal's surface are dissolved and replaced by insoluble phosphates which are rustproof. The result is a slight etching of the surface, giving a dull, nonreflecting finish.

To parkerize steels, stainless steel tanks are required, large enough to hold the solution and the metal parts. Also needed are a thermometer that will measure up to at least 180°F, measuring vessels (graduated in ounces), and a pair of tongs with at least 8-in handles.

In general, the parts to be parkerized are cleaned and degreased as described for hot caustic blueing, or old finish is removed by glass-bead blasting. After the abrasive blasting, the parts should not be handled with the bare hands as this will leave body oil on the parts and cause spotting. Use clean rubber gloves.

Prepare enough solution to cover all parts sufficiently at a concentration of about 4 oz of parkerizing solution per gallon of water. Put the solution in stainless steel tanks and bring the solution to a temperature of between 160 and 170°F. The parts are placed in the solution with tongs, taking care not to touch the parts with the bare hands. Allow the metal to react, turning the parts periodically so as to get an even treatment. The parts should not be agitated.

The metal parts are left in the solution for about 40 mintues, while the prescribed temperature is maintained. Then remove the parts with the tongs and immediately rinse in cool running water for 1 minute. Drain excess water, dry with a clean absorbent cloth, and then oil the parts.

PLATING METAL

Nickel plating has long been a favorite finish for many metal objects. It provides decorative effects, protection against rust and corrosion, and a wear-resisting surface. Although nickel has been the traditional favorite, other metals have included chromium, gold, silver, brass, and copper. Chromium, for example, when applied to metal parts provides a surface harder than the hardest steel, which protects the base metal, reduces wear, lessens friction and, at the same time, provides an attractive appearance.

Most metal plating is accomplished by a process called "electroplating" which uses an electric current to deposit the plating over the base metal. In general, the object to be plated becomes the cathode (negative plate) in an electrolyte cell that contains (in some chemical compound) the type of metal which is to be plated onto the base metal. Anodes are used on all sides of the object so that electricity may flow from all directions to the article being plated and cause an even deposit of the plated metal. See Fig. 10.3.

The exact chemicals, currents, voltages, temperatures, and procedures will vary with the kind of metal being plated and the type of method being used. For example, nickel plating often is done with an electrolyte containing nickel sulfate or nickel ammonium sulfate, to which is added ammonium sulfate to increase the conductivity, some acid to help keep the anode rough, and something like glue or glucose to make the plating extra bright.

The anodes may be of some material, such as carbon, which is not affected by the electrolytic action. When using this method, all the plated metal must come from the electrolyte and chemicals containing this metal must be added to the liquid at various intervals. In other plating methods, the anode is made of the plating metal. As an example, in plating with brass, the anode is the electrolyte which is deposited from the electrolyte onto the cathode or object to be plated. The object of this method is to get metal dissolved into the bath (electrolyte) as quickly as it plates out. As the anode metal dissolves, it generates a voltage just as dissolving a metal generates a voltage in a conventional storage battery. Under ideal conditions, this generated voltage would equal the voltage consumed in depositing metal on the cathode, so the

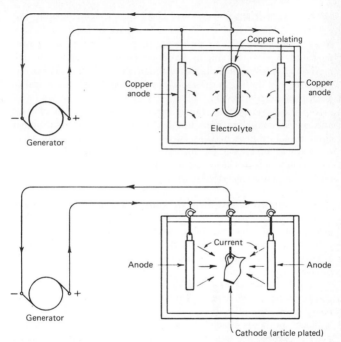

FIG. 10.3 Methods of connecting circuits for electroplating.

external source would need to provide only enough voltage to overcome the resistance in the cell and the connections.

Exact instructions are furnished with the plating kits available on the market. These should be studied carefully before using. Also, be aware that most plating kits contain highly poisonous chemicals that require certain precautions.

METAL POLISHING

Regardless of the type of finish applied, the metal must be prepared to accept the finish. Sand or bead blasting is a quick method that leaves a dull, matte finish on the surface. However, polishing and buffing is the most popular way to prepare metal to accept the desired finish.

Cloth or loose buffing wheels come in a large number of sizes and special shapes to buff practically any shape of metal object prior to coloring or plating. Most are extremely soft but when revolving at high speed (from 1700 to 3600 rpm) centrifugal force continually acts to keep them "flat" against the surface being buffed so that a relatively great amount of pressure may be applied.

Depending upon the size of the buffing wheel, it may be used in a small hand grinder (Moto-Tool), or on a large double-shafted motor, or on large revolving shafts containing many buffing wheels. The most practical size for shop use, however, is about a 1-hp double-shafted motor than can contain two buffing wheels—one on each shaft end.

What is known as a packed buffing wheel is sometimes used for color-buffing nickel, gold, and silver. It is formed by placing cardboard discs of smaller diameter in between the cloth discs.

Pieced buffs are made of remnants smaller than ordinary full discs and these are held together by continuous spiral stitching over the entire surface. This results in a wheel that is somewhat more unyielding than buffs assembled by other means.

The form of sewing on a buffing wheel is most important, as this determines the hardness of the wheel in most cases. The stitching can be spiral, crescent, radial, or other—each of which has its particular use in certain buffing applications.

The speed of the buffing wheel is also important, and Table 10.1

TABLE 10.1

Operation	Metal	Speed, ft/min
Cutting down coloring	Carbon steel	10,000 to 12,000
Cutting down coloring	Stainless steel	10,000 to 12,000
Coloring	Soft metals	5,000 to 8,000
Cutting down	Soft metals	8,000 to 10,000
Coloring	Nickel	7,000 to 8,000
Cutting down	Nickel	9,000 to 10,000
Coloring	Copper alloys	6,000 to 8,000
Cutting down	Copper alloys	8,000 to 10,000

should serve as a guide when selecting surface speeds for various applications.

In general, from Table 10-1, it can be seen that the larger-diameter wheels should operate at a slower rpm to acquire the same speed in feet per minute of the smaller wheels.

Polishing Equipment

Although it is practiced frequently in machine shops, using the lathe for polishing and grinding is not recommended. Under such conditions, fine abrasive dust will gradually find its way into the lathe bearings, and eventually will cause trouble. This dust also accumulates on the lathe ways and can scratch them.

When buffing, always wear a face mask and have adequate ventilation. It is also advisable to use a vacuum tool rigged to the wheel guard to carry off dust that forms from the buffing and polishing operation.

A pair of leather gloves is also essential to protect the hands from bad wheel burns.

The polishing head must have a motor of at least ⅓ hp for general shop use. Many prefer a 1-hp direct-drive polishing buffer. Also the polishing arms (arbors) of the power buffer must have a reasonably long shaft with the wheel mounted on its end. Otherwise it will be impossible to reach certain parts and hollows of articles to be buffed. A series of small wheels should also be mounted on the end of such shafts with a special adaptor and collet to hold them for polishing hard-to-reach areas such as trigger guards on rifles and shotguns.

The operator must be careful of sharp edges on objects being polished. A buffing wheel will rip a piece out of the hands instantly if caught just right and will sling the piece against the floor or across the room with great force.

In general buffing, the worker starts at the part of the wheel noted in a nearby illustration and then pulls the part upward against the wheel and with a pressure that will depend a great deal upon the ease with which the particular metal cuts. Very little pressure in polishing will be needed for anything but steel. For steel, considerable pressure will usually have to be used, but not all the time. For removing pits in the surface of steel, the polishing wheels cut more quickly if a circular motion is used in conjunction with moving the part upward or across the wheel. Where deep pits are involved, it is usually best to first cut these out with a belt sander using progressively finer grit sizes until the metal is smooth. Then it can be finished on the buffing wheel.

When polishing round hollow objects, such as a piece of pipe or a gun barrel, much time can be saved by using a spinning fixture. The finish will also be smoother with practically no visible abrasive marks when taken down to, say, size 500 grit buffing compound.

11

Metalworking Coolants and Lubricants

Coolants and lubricants are used in machining operations mainly to reduce heat and adhesion between the chip and tool. This cooling also prevents excessive thermal expansion and makes the work easier to handle. Furthermore, these fluids are useful in clearing metal chips away from the working area.

Intense heat produced in a metal-cutting operation will soon remove the temper from a tool point, resulting in short tool life and the production of improperly finished work. Therefore, the first requirement of a cutting fluid is cooling. Excessive friction, which is evidenced by increased heat and noticeable increase in power consumption by the tool, can be minimized by using a cutting fluid of good lubricating quality. A cutting fluid should also aid in producing a good finish and inhibit rust and corrosion. In deep boring, a washing action is normally preferred to remove excess chips.

The best type of cutting fluid for cooling is an emulsion that contains a large percentage of water. Water is seldom used alone because it is a poor lubricant and rusts the tool rapidly. However, the addition of a soluble oil to the water minimizes these effects. By increasing the amount of oil added, the solution will increase proportionately in lubricating ability. Thus, a cutting fluid has the greatest lubricating ability when it is 100 percent oil—termed a straight-oil cutting fluid. A light mineral oil alone has very little lubricating ability, so a pure or chemically treated fatty oil (animal, vegetable, or synthetic) generally is added to increase lubricating properties of a cutting fluid.

SELECTING CUTTING FLUIDS

When performing a cutting operation such as turning, threading, tapping, or milling, certain points should be considered so that the best cutting fluids will be selected.

The first consideration should be given to the metal being machined—whether it is steel, copper, cast iron, brass, or aluminum, for example. The actual machine operating conditions are also important, such as speed and the depth of cut.

Metals such as brass, cast phosphor bronze, gunmetal, and gray cast iron—which break up easily to form short chips—usually are machined without any cooling or lubricating fluid unless the speed of the operation produces an excessive amount of heat. In that case, a coolant such as a soluble oil emulsion with a high percentage of water is used. Aluminum and magnesium alloys—which are of low tensile strength—also fall into this category.

Tough metals such as steel and certain copper alloys, which form a continuous chip that presses heavily upon the face of the tool, require a lubricant as well as a coolant. A shallow cut at a low speed generally requires little coolant or lubricant; a low speed and heavy cut, particularly on a tough metal, demands a good lubricant; high speed and a shallow cut demands a coolant; high speed and heavy cut requires a cutting fluid of great lubricating as well as cooling quality. When long tool life is desired, a cutting fluid of good lubricating ability is required to reduce wear.

When cutting magnesium, an oil-and-water mixture should never be used. Magnesium, once ignited, burns more rapidly in the presence of water than in the presence of air or oil. Always use a mineral oil with a high flash point—enough to prevent ignition of the fluid. Table 11.1 gives the types of fluids preferable for common metals.

In selecting the proper cutting fluid, the composition and character of the oil is of utmost importance. Besides doing an efficient job of cooling and lubricating, the cutting fluid should not rust or corrode either the tool or the work. Corrosion is caused by plain water, excess acid or alkali, and chemical constituents that may react with the metal being worked. The cutting fluids must not decompose on standing or under operating conditions. If this decomposition occurs, impurities will form, which will cause gumming, clogging, and possibly offensive odors.

Straight mineral oils are very stable, but to improve their oiliness, fatty oils, which are not as stable and soon become rancid, must be added. Other physical characteristics to be noted are flash point and cold test. The flash point is the temperature at which oil first flashes but does not continue to burn. This should be well above any temperature at which the oil is to be used. The cold test indicates the temperature at which the fluid ceases to flow, but this latter test has very little importance in the machine shop.

The cutting fluid should be as clear as possible and free from cloudiness and supended matter when held up to the light. The color is not important, but clearness of the oil enables the operator of the machine to watch the actual cutting operation at the tip of the tool.

TABLE 11.1 Cutting Fluids to Use with Common Metals

Material	Turning speeds		Drilling speeds		Tapping speeds	
	Speed, ft/min	Lubricant	Speed, ft/min	Lubricant	Speed, ft/min	Lubricant
Aluminum	300–400	Compound or kerosene	200–330	Compound or kerosene	90–110	Kerosene and lard oil
Brass, leaded	300–700	Dry or compound	200–500	Compound	150–250	Compound or light base oil
Brass, red and yellow	150–300	Compound	75–250	Compound	60–150	Compound or light base oil
Bronze, leaded	300–700	Compound	200–500	Compound	150–250	Compound or light base oil
Bronze, phosphor	75–150	Compound	50–125	Compound	30–60	Compound or light base oil
Cast iron	50–110	Dry	100–165	Dry	70–90	Dry or Compound

Cast steel	45– 90	Compound	35– 45	Compound	20– 435	Sulfur-base oil
Copper, leaded	300–700	Compound	200–500	Compound	150–250	Light base oil
Copper, electrolytic	75–150	Compound	50–125	Compound	30– 60	Light base oil
Chrome steel	65–115	Compound	50– 65	Compound	20– 35	Sulfur-base oil
Die castings	225–350	Compound	200–330	Compound	60– 80	Kerosene and lard oil
Duralumin	275–400	Compound	250–375	Compound	90–110	Compound or kerosene and lard oil
Fiber	200–300	Dry	175–275	Dry	80–100	Dry
Machine steel	115–225	Compound	80–120	Compound	40– 70	Compound sulfur-base oil or kerosene and paraffin
Malleable iron	80–130	Dry or compound	80–100	Dry or compound	35– 70	Compound or sulfur-base oil

TABLE 11.1 *(Continued)*

Material	Turning speeds		Drilling speeds		Tapping speeds	
	Speed, ft/min	Lubricant	Speed, ft/min	Lubricant	Speed, ft/min	Lubricant
Manganese bronze	150–300	Compound	75–250	Compound	60–150	Light base oil
Manganese steel	20– 40	Compound	15– 25	Compound	10– 20	Compound or sulfur-base oil or kerosene and paraffin
Molybdynum steel	100–120	Compound	50– 65	Compound	20– 35	Sulfur-base oil
Monel metal	100–125	Compound or sulfur base	40– 55	Sulfur base	20– 30	Sulfur-base or kerosense and lard oil
Nickel silver 18%	75–150	Compound	50–125	Compound	30– 60	Sulfur-base or kerosene and lard oil

Nickel silver, leaded	150–300	Compound	75–250	Compound	60–150	Sulfur-base or kerosene and lard oil
Nickel steel	85–110	Compound or sulfur base	40–65	Sulfur-base oil	25–40	Sulfur-base oil
Plastics, hotset molded	200–600	Dry	75–300	Dry	40–54	Dry or water
Rubber, hard	200–300	Dry	175–275	Dry	80–100	Dry
Stainless steel	100–150	Sulfur base	30–45	Sulfur base	15–30	Sulfur base
Tool steel	70–130	Compound	50–65	Compound	25–40	Sulfur-base or kerosene and lard oil
Tungsten steel	70–130	Compound	50–65	Compound	20–35	Sulfur base
Vanadium steel	85–120	Compound	45–65	Sulfur base	25–40	Sulfur base

SOURCE: South Bend Lathe Co.

APPLICATION OF CUTTING FLUIDS

The method of application of a cutting fluid depends on the type of operation, but the preferred method in most cases is in a steady stream aimed directly at the exact spot where the cutting action takes place. A large stream at low velocity is preferred to a small stream at high velocity. For light work, an oil can may be used for the application, or a gravity drop-feed system, but for production work, machine tools are equipped with oil pans, pumps, and reservoirs which circulate the cutting fluid to the point of cutting. Spray-mist coolant systems use a water-base fluid which is supplied to the cutting tool by compressed air. The compressed air atomizes the cutting fluid, providing considerable cooling but little lubrication.

CUTTING FLUID CONSIDERATIONS

Machine operators must constantly guard against an unbalance between the oil and water in an emulsion cutting fluid, as troubles usually follow when the constituents of an emulsion are separated. Careless mixing and the use of impure water are among the chief causes of this problem. The manufacturer's directions for compounding should be followed explicitly. Soluble oil should mix well in water between 70 and 120°F.

An inadequate flow of cutting fluid over a hot tool point will shorten the active life of the emulsion, and therefore a generous flow is advisable. By the same token, many soluble oils break down when metals with high lead or zinc content are machined. When machining metals of this sort, use a specially compounded soluble oil.

When machining acid-pickled metals, the cutting fluid will

sometimes become excessively acid. The addition of a small amount of soda to neutralize the acid will usually solve the problem. A test may be conducted with a piece of red litmus paper. If the paper turns blue, the solution is okay; if not, add some soda ash to the oil—a little at a time—and mix well. Test with new litmus papers between applications.

Discoloring of emulsions may be caused by the collection of rust or reaction products between the metal being worked and certain constituents of the cutting fluid. Rust may be caused by improper mixing or the use of impure water. Gummy deposits may be the result of a large amount of rosin present in some cutting fluids. If a good soluble oil is used properly, little or no trouble should be encountered with bearings or the sticking of slides or guides on machine tools.

Machining troubles with straight oils are usually made evident by smoking, which is the result of exceptional surface speeds, heavy cuts on tough materials, or inadequate flow of cutting fluid. Sulfurized oils tarnish copper-rich alloys, because of a reaction that takes place in the presence of moisture. For this reason the oil should not be stored in a damp place in an open container.

APPLICATIONS OF CUTTING FLUIDS

The following applications give characteristics of the use of various cutting fluids.

Grinding

Grinding operations produce heat, resulting in an uneven temperature in the work; this causes distortion and consequent inaccuracy. The flow of a cutting fluid serves to prevent this and serves also to keep the wheel clean and free-cutting for more efficient production. Water alone will rust both the work and the machine, so the addition of a water-soluble oil (emulsion) is recommended to inhibit the corrosive action. Sufficient quantities of oil must be added to afford the necessary protection. Follow manufacturer's instructions.

Drawing, Extruding, and Pressing

These operations are accomplished by using relatively high pressure, which results in high temperature and great friction. The proper cutting fluid should provide sufficient lubrication and cooling qualities and yet not break down, decompose, or char under the severe operating conditions. It should also be easy to remove and not interfere with subsequent operations, such as welding, plating, and painting.

Dry drawing is usually done with a very dry sodium soap powder free from oil, fat, and glycerin. Wet drawing baths are merely a water solution of the sodium soap called an emulsion. Sulfonated fatty acids produce stable emulsions of fine particle size for this purpose. Graphite has been used in drawing and pressing operations, but since it causes excessive charring, its use has been restricted.

For other applications, manufacturers normally supply instructions or charts with their product giving the proper method of mixing for various applications. In using these instructions, the following details must be known:

1. Type of metal to be worked

2. Operation to be performed (threading, cutting, etc.)

3. Speed of operation

4. Depth of cut

When the above four factors are known, it is usually a simple matter to follow the instructions that accompany the water-soluble oils to obtain a correct mixture for the job at hand.

SOLID LUBRICANTS

Tools used constantly on heavy-duty production work require lubrication more frequently than others. Tools that have not been used for extended periods should be relubricated before being put back into service. In lubricating various machines, use only the recommended lubricant. An improper lubricant can result in damage to gears and mechanisms. Grease-type lubricants are often used for bearings, gears, and the like, but always follow the manufacturer's recommendations. Solid lubricants can be applied directly to the surface, or in some cases, through grease ports in the equipment.

Greases are made with a variety of soaps and are selected for various applications by the type of soap used. Two popular types are the lithium (soda-soap) grease and the modified-clay-thickened materials. For high-temperature applications, certain finely divided dyes and other synthetic thickeners are applied.

Greases also vary in volatility and viscosity according to the oil used. Since volitility can affect the useful life of the bulk applied to the bearing and viscosity can affect the load-carrying capacity of the grease, they must both be considered in selecting a grease.

When used on gears and certain bearings, many greases are thickened with carbon, graphite, molybdenum disulfide, lead, or zinc oxide. These additives can also be used by themselves for providing dry lubrication during high-load, slow-speed load conditions.

Most requirements for bearing lubrication can be met by three viscosities of machine oil and two grades of cup or ball-bearing grease. In some cases, the addition of graphite to oil in a multiple-spindle gear box lowers the temperature 15°F in less than an hour. The application can also reduce the current input slightly, or about 1 amp and stops foaming of the oil.

A mixture of graphite and a little lubricating oil, brushed on ways, gibs, and other bearing surfaces, reduces chatter caused by momentary breaking down of oil films. Graphite is commonly used in new and refitted bearings during the running-in period, on gears and in bearing grease, and in new machines to reduce static or starting friction.

12

Bench Work and Tools

"Bench work" is the term used when the machinist works at the bench with hand tools rather than with machine tools, and involves such operations as laying out, filing, hand sawing, chipping, deburring, hand reaming, scraping, hand tapping, bench assembly, hand spring winding, hand honing and polishing, and all of the many shop jobs done at the bench or in a bench vise.

The main requirements for bench work are a good solid workbench and a large heavy machinist's vise; a smaller vise (for small workpieces) is also advisable. The bench may be constructed of wood or metal and should be of a height convenient for the persons working at it. The exact height will therefore depend upon the height of the user and how far the vise jaws project above the table or bench top. For the average person, however, the height will be around 30 to 32 in. Furthermore, the bench top must be solid, as the best work cannot be obtained from a shaky workbench.

Good lighting is also necessary for good work. Therefore, the bench should be well-lighted with fluorescent lighting fixtures to produce at least 100 footcandles of illumination directly on the bench without undue shadow. Some workers prefer 150 to even 200 footcandles of light, plus an adjustable direct light for individual work.

LAYOUT WORK

Before most objects can be machined in machine-shop power tools, the exact dimensions, lines, angles, and the like must be determined beforehand and then scribed on the workpiece for the guidance of machine operators. This operation is termed "layout work" in the machine shop. In all cases, much care must be used to ensure an accurate layout, using the measuring tools described in Chap. 6; all lines and center points should be exact and sharp.

In laying out work for the machine operator, many lines are scribed directly onto bare metal. However, these markings will appear sharper if the metal is first coated with some substance or compound. If precise dimensions are not required, chalk carefully rubbed on the surface of the workpiece prior to scribing the lines will usually suffice. However, for fine, exact layouts, a special marking solution is recommended. The two most common ones are copper sulfate solution and blue layout dye.

When copper sulfate is applied to a steel surface, a dull coppered surface appears which is ideal for scribing lines and centers. The blue layout dye leaves a blue or blue-black color to the surface and is the common surface treatment throughout the industry. When using either of these solutions, be sure to flush the surfaces with warm water after the machining is complete, and then treat the surface with oil, as both solutions will promote rapid rust.

If extremely accurate results are to be obtained in laying out flat work, special plates are used on which the work and tools are placed. These are known as leveling, surface, or laying-out plates. When such a plate is not immediately available, a heavy piece of mirror may be used, but of course the glass is subject to breakage, and much care must be used. The work may be placed directly on the surface of the plate or held upon parallel bars or blocks placed on the plate. In other cases, it is convenient to clamp the work to knee or angle irons, which are then placed on the surface plate. These plates—such as those offered by Starrett—are made from granite blocks especially selected for this purpose or from well-seasoned cast iron. See Fig. 12.1.

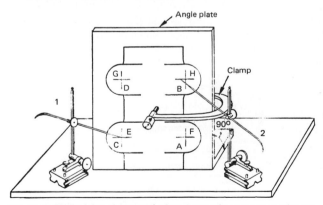

FIG. 12.1 Typical flat plates used when extremely accurate results are to be obtained in layout work.

BENCH TOOLS

Bench Vise

While there are several types of bench vises in use today, most machinists are using what is known as a combination vise, which has inner teeth built as an integral part of the vise for holding pipe and round stock. A standard machinist's vise is also frequently used, and is available in plain, self-adjusting, quick-acting, and swivel types. The main purpose of all bench vises is to hold pieces of work firmly. See Fig. 12.2.

Hammers

Hammers will find a variety of uses in bench work: driving pins, peening, riveting, stamping, and other operations requiring a sharp blow.

The no-mar type of hammer is also used on the bench. These nylon, rubber, or leather-headed hammers may be used for a multitude of jobs around the shop where it is necessary to avoid mar-

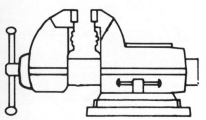

FIG. 12.2 Machinist bench vise.

ring the work while pounding. Brass hammers also fit into this category, and when a combination hammer is used, it makes an excellent all-purpose nonmarring hammer for the machinist. This combination hammer has a brass hammer tip on one end of the head, and a nylon or phenolic tip on the other.

Metalworking hammers may be classified with respect to peen as

1. Ballpeen
2. Straightpeen
3. Crosspeen

These hammers (Fig. 12.3) are used to indent or compress metal and to expand or stretch metal. Marred metal areas can sometimes be peened back into their original condition if care is taken. A bar or shaft may be straighted by peening with the convex portion of the hammer.

Screwdrivers

The selection of screwdrivers seldom receives the care needed in most machine shops. The effective holding power of a screwdriver depends upon the quality of steel in its blade, the design of the blade, and the external force that may be applied to the screwhead. The blade should also be fitted to the width of the slot for best results.

The double-wedge-type screwdriver is the most common around the machine shop. With this type of screwdriver, the blade transmits its torque to the top of the screw slot, and when used on small screws, there is a good possibility that the screw will be scored or worse yet, cause one section of the screwhead to break off if seated tightly. Wedge-shape tips (Fig. 12.4) also tend to back

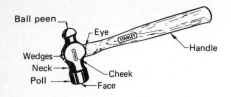

Ball peen — Eye
Wedges — Neck — Cheek
Poll — Face — Handle

Straight peen

Cross peen — Eye
Wedges — Neck — Cheek
Poll — Face — Handle

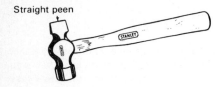

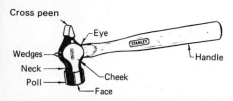

FIG. 12.3 Three types of hammers with respect to peen. (*a*) Ball peen, (*b*) straight peen, and (*c*) cross peen.

the driver out of the screw slot, again causing damage to the screwhead.

A screwdriver tip, ground to properly fill a screw slot, is the best type to use for most machine-shop work—especially when used on precison work such as toolmaking, firearm assembly, and the

 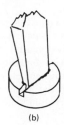

FIG. 12.4 Wedge-shape screwdriver blade in use. (*a*) Note lack of close fit—the blade transmits its torque only to the top of the screw slot. (*b*) This screw slot was damaged by improper blade fit.

like. See Fig. 12.5. With this type of screwdriver, the torque is applied at the bottom of the slot where the screw is the strongest; also, the blade will fill the slot completely (and should be the same width as the shank).

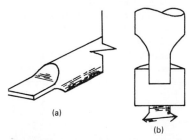

FIG. 12.5 A screwdriver tip properly ground to fill a screw slot. (*a*) This particular blade is ideal for gun work. It has been cut to conform perfectly to a screw of a particular dimension. (*b*) Notice that this screwdriver blade fits the slot perfectly.

Cold Chisel

Various cold chisels are used in the machine shop, the most common being the flat cold chisel. This type is used for cutting and chipping and for the removal of rivet and nut heads.

The diamond point chisel is used for chipping v-shaped oil grooves and sharp corners, while the cape chisel is forged to produce a cape or flare for the widest flat at the cutting edge. This latter type is used for cutting narrow slots, keyways, and rectangular grooves.

The round-nose chisel is used for producing oil grooves and other concave surfaces, and can be used for drawing a drill back to the true location of the hole to be drilled.

Filing

An expert machinist with a set of files and cold chisels can perform practically all machine-shop operations, for with just these tools, other tools can be built to build new tools, and so forth. While hand filing is being used less and less in the average machine shop, every machinist should be able to select and use files for the job at hand. Therefore, a complete chapter (Chap. 28) has been devoted to this subject, and no further instructions are given here.

Polishing

Where a particularly smooth surface is required, or where a brilliant finish is desired, the surfaces are polished with abrasive cloth or paper made by securing fine abrasive to a cloth or paper backing. These abrasive cloths are furnished with a large variety of grit sizes and abrasive materials for use on a wide range of applications. With enough work, these abrasive cloths are capable of removing large amounts of material. However, they are generally used for conditioning, cleaning, or polishing operations that require little material removal.

Tapping

Taps are used to cut internal threads. A standard hand tap set consists of three taps:

1. A taper or starting tap
2. A plug tap
3. A bottoming tap

Each has a different starting chamfer or end taper. Each, however, has the same diameter.

Before a tap can be used, a hole must be drilled in the workpiece with a tap drill of the correct size. Complete details of tapping can be found in Chap. 15.

Sawing

The best hacksaw frame is not expensive, so only the best should be purchased. It will be used a great deal around the shop. Likewise, only the best tungsten high-speed hacksaw blades can be expected to hold up for all kinds of machine shop work. A quality hacksaw frame will help eliminate blade twist or wobble. On the better frames, the blade may be inserted in a conventional manner for cutting stock down to the surface and may also be inserted at a 90 degree angle for cutting flush with the surface.

The choice of teeth per inch on the blade will depend on the type of work the saw is to be used on. In general, the thinner the material to be cut, the higher number of teeth. Soft material requires fewer teeth per inch than hard material.

When sawing, the workpiece must be held securely in a vise and positioned so that the cutoff is near the vise jaws to prevent shattering. To start a cut, use the thumb as a guide and saw slowly with short strokes until the cut is well started (Fig. 12.6). As the cut deepens, grip the front end of the frame firmly and take a full-length stroke.

Face the work while sawing with one foot in front of the other and approximately 12 in apart. Pressure should be applied on the forward stroke only because the blade only cuts on the forward stroke. On the backstroke do not let the teeth drag, as this will dull the teeth and may cause the blade to break. See Fig. 12.7.

Reaming

Reaming is the operation of finishing a hole to exact size and to an acceptable degree of smoothness. The practice is necessary on

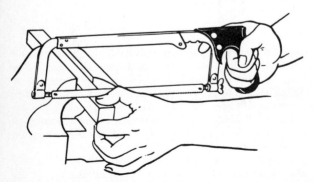

FIG. 12.6 Starting a cut with a hacksaw.

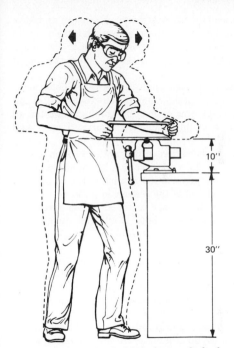

FIG. 12.7 Do not let the blade drag on the back-ward stroke as this will tend to dull the teeth.

some applications because a drill bit will not produce a hole to the exact dimensions of the drill bit. Therefore, when a precise hole must be produced, it is first drilled a little under size, and then reamed to the correct size and smoothness.

Hand reamers are usually either straight or tapered and have either straight or helical flutes. In use, a tap wrench is used to hold the reamer by the square end. The opposite end of the reamer is slightly tapered for a short distance to ensure square starting of the reamer in the hole.

The tapered reamers are used for finishing taper holes accurately and smoothly. The tapered reamers sometimes utilize spiral flutes which have a shearing action and eliminate chatter. all standard tapers are available including Morse, Brown and Sharpe, and taper pins.

Holes to be reamed must be first drilled not more than .005 to .010 in under size. Clamp the workpiece in a bench vise with the hole vertical. Attach an adjustable tap wrench to the square end of the reamer and set the end of the reamer into the hole. Brush a clean cutting oil over the reamer teeth, and then rotate the reamer clockwise, making sure to hold it at right angle to the hole. If the hole is the correct size, the reamer will soon align itself.

At first, a slight downward pressure may be needed, but a hand reamer will practically feed itself once it is started. It may advance into the hole as much as one-fourth the reamer diameter for each revolution. Never turn a reamer backward, as this will quickly dull the delicate cutting edges.

During the cutting operation, always make certain that the reamer is kept cutting. To do otherwise will invite chatter, which causes a series of rough lines on the cutting surface. Once started, the chatter always gets worse, so at the first signs of chatter, increase the downward pressure slightly on the reamer to keep it cutting.

Continue to turn the reamer clockwise until the hole is reamed, and then continue to turn the reamer clockwise as it is backed out of the hole.

HAND BROACHING

After a hole has been reamed, it is often necessary to cut a keyway to a standard depth and width. This can be done at the bench with a set of broaching tools. The set usually consists of cutting tools and bushings for use in various size holes.

To broach a keyway by hand, first select the correct bushing for the hole in question, and insert the bushing in the work. Insert the broach of the correct size for the keyway being cut. Place the workpiece with bushing and broach in an arbor press, lubricate, and push the broach through the workpiece. In doing so, make certain that the chips are kept clean by brushing away from the teeth. Remove the broach and measure depth of cut. If additional depth is needed, a shim can be placed behind the broach, or a different size broach can be used.

HAND SCRAPERS

Scrapers are used to make accurate surfaces. They remove only a very small amount of material. Hand scraping, however, requires a good deal of skill to remove high spots from the surface of the work. Common applications include scraping the beds of lathes that are being reconditioned to ensure an accurate flat surface.

The most common scraper used in the machine shop is the flat-type scraper. The bearing or half-round scraper is a slender tool made of hardened steel especially shaped and curved. It is used for scraping a bearing surface so that a shaft will fit into it properly. The three-square scraper is a hardened tool used to remove burrs and sharp internal edges from bushings.

To use a flat scraper, first clear the work and surface plate with a solvent such as AWA 1,1,1, and then apply a thin layer of prussian blue to the surface plate. Place the work to be scraped face down on the surface plate and move it in a figure-eight pattern. High spots will show in blue and must be removed by scraping, which is done by pushing the cutting edge of the scraper across the high spots. Cutting is done on the push stroke. Only a small area should be covered at a time. Repeat the process as often as necessary, covering with prussian blue until the surface is completely flat.

Frosting or flaking is the process of decorating a scraped surface and is also done with a handscraper. Light cuts are made, each in a different direction until the entire surface is decorated. Very short strokes of approximately ¼ to ½ in are common. This process results in a finished surface that is attractive in appearance.

13

Principles of Lathe Operation

The screw-cutting metal-turning lathe is the oldest and most important machine tool, and from it practically all other machine tools have been developed. Lathes vary in size from the small jeweler's or clockmaker's lathes for making very fine, miniature parts to the large gap-bed lathes and special-purpose lathes used in high-production work. In between these extremes are many models of varying lengths and capacities.

The maximum size—that is, the diameter and length—of work that can be handled by the lathe is used to designate the size of the lathe. For example, a 9 × 36 in lathe is one having a swing over the bed sufficient to take work up to 9 in in diameter and a distance between lathe centers of 36 in.

When selecting a lathe for machine-shop work, careful consideration must be given to the size and amount of work that the lathe will be required to handle. Ideally, the lathe selected should have a swing capacity and distance between centers at least 10 percent greater than the largest job that is anticipated.

PARTS OF A LATHE

A lathe is made up of many parts. The principal parts are shown in Fig. 13.1 and include bed, head stock, tail stock, carriage, feed mechanism, thread-cutting mechanism, and others.

Bed

The lathe bed is the foundation on which the lathe is built, so it must be substantially constructed and scientifically designed. The two types in common use are flat and prismatic v-ways. Flat ways are shown in Fig. 13-1.

Head Stock

The head stock is one of the most important parts of the lathe and should be back-geared for more versatility. The back gears provide a means of controlling the spindle speed. A lever engages the back gear for slow spindle speeds (75 to 280 rpm) or disengages it for high spindle speeds from about 300 to 1100 rpm or more.

Tail Stock

The tail stock assembly is movable on the bed ways, and carries the tail stock spindle. The tail stock spindle has a standard Morse taper, in most cases, to receive various types of lathe centers. The tailstock handwheel is at the other end to give longitudinal movement when the workpiece is mounted between centers.

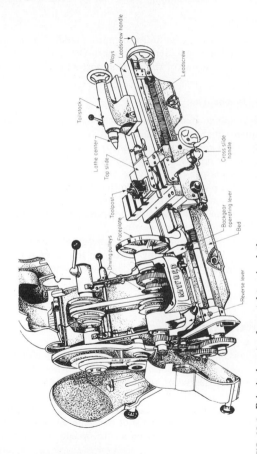

FIG. 13.1 Principal parts of a metal-turning lathe.

Carriage

The lathe carriage includes the apron, saddle, compound rest and tool post. Since the carriage supports the cutting tool and controls its action, it is one of the most important units of the lathe.

Feed Mechanism

Quick-change gear lathes are preferred in machine shops where frequent changes of threads and feeds are required. On most lathes, the quick-change gearbox is located directly below the head stock on the front of the lathe bed. A wide range of feeds and threads per inch may be selected by positioning the gears. The index plate is an index to the lever settings required to position the gears for the different feeds, and inches that the carriage will move per revolution of the spindle is given in each block for the corresponding gear setting.

The reversing lever is used to reverse the direction of rotation of the screw for chasing right- or left-hand threads, and for reversing the direction of feed of the carriage assembly. Levers on the quick-change gearbox should never be forced into position.

INSTALLATION

The installation of a lathe should be carefully planned, following certain essential guidelines during the process. Doing so will allow the highest degree of accuracy with the least amount of repairs and maintenance.

One consideration that is often overlooked is the preparation of facilities prior to the arrival of the lathe. Even smaller lathes will be shipped in crates having a total weight of 500 lb or more. In most cases, the buyer will be notified as to the date of shipment and approximate time of arrival. When this information is received, immediate arrangements should be made to have help on hand to unload the crates and also provide adequate storage facilities until the lathe is uncrated and set up. If proper tools are not available at the delivery location, consider renting or borrowing such items as pinch bars, dollies, forklifts, chain hoists, ropes, and pallets, and hiring extra personnel to facilitate handling and setting the lathe in place.

Consideration should also be given to the area in which the lathe will be housed. Specifications containing the exact dimensions, weight, and the like should be obtained from the manufacturer prior to the arrival of the lathe so that a suitable area can be prepared in advance of the lathe's arrival. In selecting a suitable site, sufficient clearance on all sides of the lathe should be provided for the motor and for access to the geers, change wheels, and other controls. Sufficient room should also be provided behind the lathe for cleaning, oiling, and attaching accessories such as a taper turning attachment. The tail stock end will require room to remove the tail stock from the ways, to operate the handwheels, and to perform other operations.

Most lathe manufacturers use great care in packing their products to ensure that the user will receive them in perfect condition. The unpacking should be carried out with the same care to avoid possible damage. As each item is unpacked, check it over very carefully for damage, and if none is found, set it aside and check the item off the master list. If any damage of shortages are found, notify both the shipper and the shipping company immediately. Any great delay might cause problems.

Many lathe buyers have the tendency to destroy all packing material as it is taken out of the box or crate. This should be avoided. Rather, all loose packing material (such as wood wool) should be set aside and thoroughly searched after everything is unpacked to ensure no items of value will be destroyed; this is especially true if the inventory turns up short. It is also advisable to hang on to the packing material, boxes and crates in case any of the items have to be returned.

Most lathes and their accessories are shippped from the factory with all parts protected by a rust preventive. This substance must be removed before the items are put to use and before it gets on clothing and other articles. Remove the rust preventive as soon as possible from all parts—especially those which will be handled and assembled first. To remove it, a cleaning solution such as AWA 1,1,1 or gasoline is preferred. However, if gasoline is used, be extremely careful not to let any source of heat near it. Fires and explosions are often the result of improper use.

Cleaning solution applied to a rag is often all that is necessary to get the rust preventive removed completely. However, cotton swabs are excellent for getting into tight places such as between gear teeth and in the various recesses fuond in every lathe. Immediately after removing the rust preventive, apply a thin coat of oil such as SAE 30 motor oil, or spray the parts with WD-40. This latter oil is especially handy for hard-to-reach areas.

The next order of business to assure good work and efficient performance is to set up the lathe at a suitable height for ease of working. For the average person, standard work height is about 33 in. Then the lathe must be leveled carefully on a solid floor or bench, depending on whether it is a floor model or a bench type. Use an accurate carpenter's level or a precision-ground machinist's level. First place it crosswise on the ways close to the head stock, then crosswise at the extreme tail stock end, and longitudinally about midway on either front or rear way. Place thin shims of hardwood or metal under the legs until the required level is attained. Then bolt the legs to the floor or bench, gradually tightening the nuts in sequence a little at a time until the lathe is anchored firmly.

CENTERING WORK

Both centers of a lathe must be in perfect alignment. One way to check center alignment is to use a test bar such as the one shown in Fig. 13.2. This can be a 12-in length of round stock, 1 in in diameter, and must be concentric and straight to within .001 in. By using a bar of this type, the centers can be aligned easily.

One favorite method used to align lathe centers is the "trial cut" method where a light cut is made on the workpiece. A measurement is then taken at the tail stock end of the workpiece and at a point midway between centers. If there is a difference in the two readings, then the tail stock is moved an amount equal to the

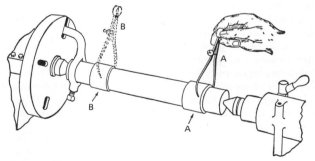

FIG. 13.2 Test bar for use in checking lathe alignment.

difference. If the tail stock end is the larger of the measurements, the tail stock is moved toward the operator; if the tail stock end is smaller, it is moved away from the operator.

To test live centers for concentricity, use a dial indicator placed in the tool post or on the carriage, and with the indicator contact point touching the center, rotate the spindle slowly by hand—at the same time noting any movement on the dial. If the dial indicates an eccentricity in excess of the allowable limits for the job, the cone point should be machined true.

After the lathe has been tested and put in good working order, it is ready for turning. First, accurately located and drilled center holes are required in both ends of work that is to be held between centers in the lathe. These holes, which serve as bearing points for the lathe centers, are made to the standard angle of 60 degrees, and the countersunk or tapered portion of all center holes must have the same angle for a perfect fit, as shown in Fig. 13.3.

When the center-drilled work is being mounted between the lathe centers, the countersunk holes must be kept free of dirt and chips to assure accurate centering, and the tail stock center must be oiled well to lessen friction. While there is no standard size for center holes, the following chart may be used as a guide for work up to 4 in in diameter and over (Fig. 13.4). Methods of locating centers in work are described in Chap. 6.

Before mounting work between centers, check the centers for alignment, as shown in Fig. 13.5. If the tail stock center does not line up, loosen the tail stock clamp bolt and set over the tail stock top in the proper direction by adjusting the tail stock setover screws as shown in Fig. 13.6.

Place a drop of oil in the center hole for the tail stock center point before mounting the work between centers. The tail of the lathe dog should fit freely into the slot of the face place so that the work rests firmly on both the head stock center and the tail

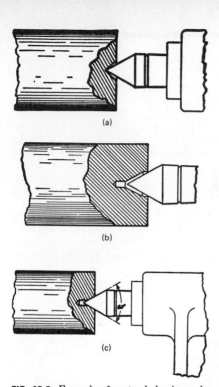

FIG. 13.3 Example of center holes in workpiece which are used as bearing points for the lathe centers. (*a*) This center hole had been poorly drilled; it is too shallow and at an incorrect angle. (*b*) This center hole is improper; it is drilled too deeply to fit the center. (*c*) Here is a correctly drilled and countersunk center hold; it fits the lathe center perfectly.

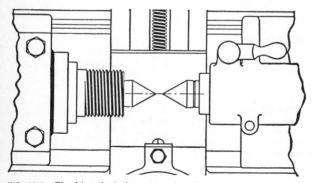

	Diameter of work, W	Large diameter of counter-sunk hole, C	Diameter of drill, D	Diameter of body, F
	3/16 to 5/16 in	1/8 in	5/64 in	3/16 in
	3/8 to 1 in	3/16 in	7/64 in	1/4 in
	1 1/4 to 2 in	1/4 in	1/8 in	5/16 in
	2 1/4 to 4 in	5/16 in	3/16 in	7/16 in

FIG. 13.4 Size of center holes for work up to 4 in in diameter.

stock center, as shown in Fig. 13.7. Make sure that the lathe dog does not bind in the slot of the face plate.

When turning work, the cutting edge of the cutter bit and the end of the tool holder should not extend over the edge of the compound rest any farther than necessary. The tool should also be set so that if it should happen to slip, it will not dig into the work, but should move away from the work.

FIG. 13.5 Checking the lathe centers for alignment.

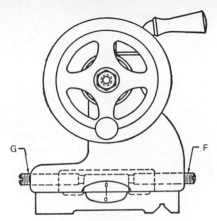

FIG. 13.6 Adjustment of tail-stock setover screws.

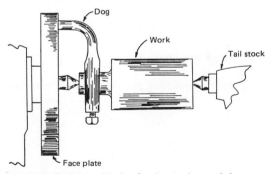

FIG. 13.7 Position of lathe dog in turning work between centers.

The feed of the tool should be toward the head stock, if possible, so that the pressure of the cut is on the head spindle center which revolves with the work.

The rate of the power feed depends on the size of the lathe, the nature of the work, and the amount of stock to be removed. On a small lathe, a feed of .008 in per revolution of the spindle may be used, but on larger sizes of lathes, feeds as coarse as .020 in per revolution are often used for rough turning. Care must be taken when turning long slender shafts, as a heavy cut with a coarse feed may bend the shaft and ruin the work.

The most efficient cutting speed for turning varies with the kind of metal being machined, the depth of the cut, the feed, and the type of cutter bit. If too slow a cutting speed is used, much time may be lost, and if too high a speed is used the tool will dull quickly. The chart in Fig. 13.8 gives cutting speeds recommended for high-speed steel cutter bits. however, if a cutting lubricant is used, these speeds may be increased by as much as 50 percent; if tungsten-carbide–tipped cutting tools are used, these speeds may be increased from 100 to as much as 800 percent.

SECURING WORKPIECE IN LATHE

Work is held in a lathe for machining usually by one of the following techniques:

1. Between centers
2. Independent chuck
3. Universal chuck
4. Collets
5. Faceplate

Cutting Speeds for Turning—Drilling—Tapping With High Speed Steel Cutting Tools

Material	Turning Speeds		Drilling Speeds		Tapping Speeds	
	Ft. per Minute	Lubricant	Ft. per Minute	Lubricant	Ft. per Minute	Lubricant
Aluminum	300-400	Comp. or Kerosene	200-330	Comp. or Kerosene	90-110	Kerosene & Lard Oil
Brass, leaded	300-700	Dry or Comp.	200-500	Comp.	150-250	Comp. or Lt. Base Oil
Brass, red and yellow	150-300	Comp.	75-250	Comp.	60-150	Comp. or Lt Base Oil
Bronze, leaded	300-700	Comp.	200-500	Comp.	150-250	Comp. or Lt. Base Oil
Bronze, phosphor	75-150	Comp.	50-125	Comp.	30- 60	Comp. or Lt. Base Oil
Cast Iron	50-110	Dry	100-165	Dry	70- 90	Dry or Comp.
Cast Steel	45- 90	Comp.	35- 45	Comp.	20- 35	Sul. Base Oil
Copper, leaded	300-700	Comp.	200-500	Comp.	150-250	Lt. Base Oil
Copper, electro.	75-150	Comp.	50-125	Comp.	30- 60	Lt. Base Oil
Chrome Steel	65-115	Comp.	50- 65	Comp.	20- 35	Sul. Base Oil
Die Castings	225-350	Compound	200-330	Compound	60- 80	Kerosene & Lard Oil
Duralumin	275-400	Compound	250-375	Compound	90-110	Comp. or Ker. and Lard Oil
Fiber	200-300	Dry	175-275	Dry	80-100	Dry
Machine Steel	115-225	Compound	80-120	Compound	40- 70	Comp., Sul. Base Oil or Kero. & Para
Malleable Iron	80-130	Dry or Comp.	80-100	Dry or Comp.	35- 70	Comp. or Sul. Base Oil
Mang. Bronze	150-300	Comp.	75-250	Comp.	60-150	Lt. Base Oil
Mang. Steel	20- 40	Comp.	15- 25	Comp.	10- 20	Comp. or Sul. Base Oil or Ker. & Para
Moly. Steel	100-120	Comp.	50- 65	Comp.	20- 35	Sul. Base Oil
Monel Metal	100-125	Comp. or Sul. Base	40- 55	Sul. Base	20- 30	Sul.Base or Kero. and Lard Oil
Nickel Silver 18%	75-150	Comp.	50-125	Comp.	30- 60	Sul.Base or Kero. and Lard Oil
Nickel Silver, leaded	150-300	Comp.	75-250	Comp.	60-150	Sul.Base or Kero. and Lard Oil
Nickel Steel	85-110	Comp. or Sul. Base	40- 65	Sul. Base Oil	25- 40	Sul. Base Oil
Plastics, hot-set molded	200-600	Dry	75-300	Dry	40- 54	Dry or Water
Rubber, Hard	200-300	Dry	175-275	Dry	80-100	Dry
Stainless Steel	100-150	Sul. Base	30- 45	Sul. Base	15- 30	Sul. Base
Tool Steel	70-130	Comp.	50- 65	Comp.	25- 40	Sul.Base or Kero. and Lard Oil
Tungsten Steel	70-130	Comp.	50- 65	Comp.	20- 35	Sul. Base
Vanadium Steel	85-120	Comp.	45- 65	Sul. Base	25- 40	Sul. Base

The above speeds have been collected from several sources and are suggested as practical for average work. Special conditions may necessitate the use of higher or lower speeds for maximum efficiency.

FIG. 13.8 Table of cutting speeds for high-speed steel cutter bits. (*South Bend Lathe Co.*)

LATHE CUTTING TOOLS

Cutting tools used for machining in metal-turning lathes are numerous. A few are shown in Fig. 13.9. To machine metal accurately and efficiently, it is necessary to have the correct type of lathe tool with a sharp, well-supported cutting edge (ground for the particular job at hand) and set at the correct height. The ones shown in the illustration are the most common in use, and a few of them are described as follows:

Cutter bit for rough turning.   Used for removing a large amount of material in a short period of time.

Side tools.   Used for longitudinal and transverse turning and for turning acute corners.

Planing tool.   Used to obtain a smooth transverse surface.

Parting-off tool.   Used for grooving and parting-off workpieces. In using this type of bit, the tool bit point must be mounted to the exact center height (or just slightly higher) and the slowest speeds must be used. Also use lubrication during the cutting operation.

Thread-cutting tool.   Used for cutting screw threads. When using this type of cutter bit to cut screw threads, always keep the work flooded with oil to obtain a smooth thread. Machine oil may be used, but special thread-cutting oil is better.

Inside boring tool.   Used for turning inside work.

SPECIAL LATHE WORK AND TOOLS

There are also classes of work involved in machine-shop activities that require special tools; the most important are described below.

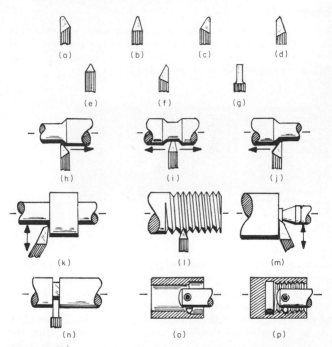

FIG. 13.9 Some cutting tools used for metal turning. (*a*) Left-hand turning tool, (*b*) round-nose turning tool, (*c*) left-hand facing tool, (*e*) threading tool, (*f*) right-hand facing tool, (*g*) cutoff tool, (*h*) left-hand turning tool, (*i*) round-nose turning tool, (*j*) right-hand turning tool, (*k*) left-hand facing tool, (*l*) threading tool, (*m*) right-hand facing tool, (*n*) cutoff tool, (*o*) boring tool, and (*p*) inside threading.

Knurling

Knurling is the process of embossing the surface of a piece of work in the lathe with a knurling tool secured in the tool post of the lathe. The pattern of the knurl is always the same (Fig. 13.10), but is of different grades—coarse, medium, or fine.

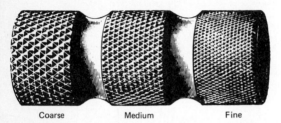

Coarse   Medium   Fine

FIG. 13.10 Coarse, medium, and fine knurling patterns.

For all knurling operations, the lathe should be arranged for the slowest back-geared speed. After starting the lathe, force the knurling tool slowly into the work at the right end until the knurl reaches a depth of about ¹⁄₆₄ in. Then engage the longitudinal feed of the carriage and let the knurling tool feed across the face of the work. Plenty of oil should be used on the work during this operation.

When the left end of the knurl roller has reached the end of the work, reverse the lathe spindle and let the knurling tool feed back to the starting point. Do not remove the knurling tool from the impression but force it into the work another ¹⁄₆₄ in, and let it feed back across the face of the work. Repeat this operation until the knurling is finished.

Filing and Polishing

All tool marks can be removed from metal parts and a smooth, bright finish obtained on the surface by filing and polishing. Use a fine mill file and file with the lathe running at a speed so that the work will make two or three revolutions for each stroke of the file. File just enough to obtain a smooth surface. If too much filing is done, the work will be uneven and inaccurate.

Keep the left elbow high and the sleeves tightly rolled up so there will be no danger from being dragged into the headstock or workpiece as it revolves. Also keep the file clean and free from chips; use a file card frequently.

Once the filing has been completed, a very smooth, bright finish may be obtained by polishing the work with several grades of emery cloth. Use oil on the emery cloth and run the lathe at high speed. Be careful not to let the emery cloth wrap around the revolving work.

Lapping

Certain hardened parts are often finished in the lathe by lapping. Emery cloth, emery dust and oil, diamond dust, and other abrasives are used. Usually the lathe spindle is operated at high speed.

The lap may be very simple—consisting of a strip of emery cloth attached to a shaft—or it may be elaborately constructed of lead, copper, etc. Some very fine and precise work may be accomplished by careful lapping.

Spring Winding

Coil springs of all kinds may be wound on the lathe. Special mandrels are used for irregularly shaped springs. The lead screw and half nuts of the lathe are usually used to obtain a uniform lead so that the coils are all equally spaced.

The Use of the Center Rest:

The center rest (Fig. 13.11) is used for turning long shafts and for boring and threading spindles. To mount work in the center rest, first place the center rest on the lathe, then place the work between centers, slide the center rest to its proper position, and adjust the jaws upon the work. Careful adjustment is required because the work must revolve in these jaws. When the jaws are adjusted properly so that the work revolves freely, clamp the jaws in position, fasten the work to the head spindle of the lathe and slide the tailstock out of the way. One end of the work may be held in a chuck, but for fine, accurate work, the chuck should not be used. Rather, work should be mounted on centers, using a lathe dog.

The Use of the Follower Rest

The follower rest is attached to the saddle of the lathe to support work of small diameter that is liable to spring away from the cutting tool. The adjustable jaws of the follower rest bear directly on the finished diameter of the work. As the tool feeds along the work, the follower rest (being attached to the saddle) travels with the tool.

Manufacturing Duplicate Parts in the Lathe

The modern back-geared screw-cutting lathe can be fitted with rapid production attachments and used to advantage on many

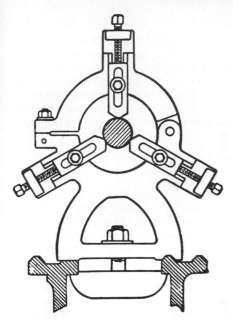

FIG. 13.11 The center rest acts as a support for long workpieces.

manufacturing operations. The accuracy of the lathe combined with the efficiency of the special attachments makes it ideal for production work requiring special accuracy.

When the lathe is equipped with special tools, it serves as a special machine. Then when the job is finished, the special tools can be removed and the lathe used for regular work.

Milling in the Lathe

A milling and keyway-cutting atttachment will take care of a great deal of milling in the shop that does not have enough volume to warrant the installation of a milling machine. The cut is controlled by the handwheel of the lathe carriage with the cross-feed screw of the lathe and the vertical adjusting screw at the top of the milling attachment. All milling cuts should generally be taken with the rotation of the cutter against the direction of the feed.

TAPER TURNING

Most lathe head stock and tail stock spindles are designed to take tapered centers, most of which are reamed to standard Morse tapers running from no. 0 through no. 7. The angle of these tapers (taper per foot) varies slightly through the whole range.

Figure 13.12 shows a cross-sectional view of a no. 2 taper with dimensions of tapers from no. 0 through no. 7, Fig. 13.12b.

There are various ways of turning a tapered piece in the lathe: for long work, where the angle is slight, good results are obtained by setting the tail stock off center. However, when the work is short, and the taper steep, this method is not the most accurate. Nor can this method be used for boring tapered holes.

For short tapers of any angle, and for tapered holes, an effective method is to chuck the work and use the compound rest, set at the angle required. A third and perhaps better way of turning tapers is to use a taper attachment. This attachment can be set, either for turning or for boring, to specific angles or to any desired taper per foot. Duplicate work can be turned accurately to the proper taper, whatever the length of the piece, if within the range of the attachment.

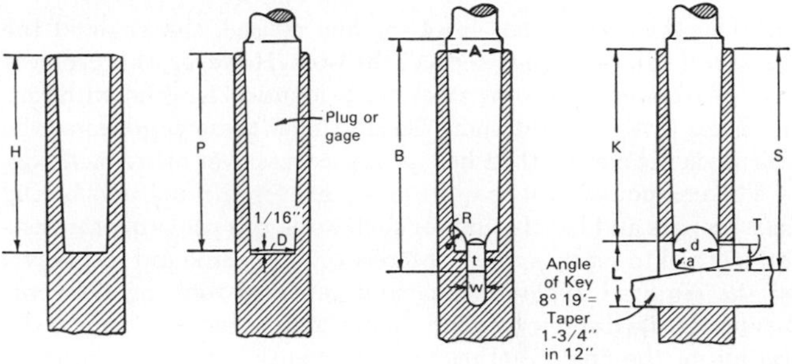

(a)

Number of Taper	Diam. of Plug at Small End, Inches	Diam. at End of Socket, Inches	SHANK: Whole Length of Shank, Inches	SHANK: Shank Depth, Inches	Depth of Hole, Inches	Standard Plug Depth, Inches	TONGUE: Thickness of Tongue, Inches	TONGUE: Length of Tongue, Inches	TONGUE: Rad. of Mill for Tongue, Inches	TONGUE: Diameter of Tongue, Inches	TONGUE: Radius of Tongue, Inches	KEYWAY: Width of Keyway, Inches	KEYWAY: Length of Keyway, Inches	KEYWAY: End of Socket to Keyway, Inches	Taper per Foot	Taper per Inch	Number of Key
	D	A	B	S	H	P	t	T	R	d	a	W	L	K			
0	.252	.3561	2 11/32	2 7/32	2 1/32	2	5/32	1/4	5/32	.235	.04	.160	9/16	1 15/16	.62460	.05205	0
1	.369	.475	2 9/16	2 7/16	2 3/16	2 1/8	13/64	3/8	3/16	.343	.05	.213	3/4	2 1/16	.59858	.04988	1
2	.572	.700	3 1/8	2 15/16	2 5/8	2 9/16	1/4	7/16	1/4	17/32	.06	.260	7/8	2 1/2	.59941	.04995	2
3	.778	.938	3 7/8	3 11/16	3 1/4	3 3/16	5/16	9/16	9/32	23/32	.08	.322	1 3/16	3 1/16	.60235	.05019	3
4	1.020	1.231	4 7/8	4 5/8	4 1/8	4 1/16	15/32	5/8	5/16	31/32	.10	.478	1 1/4	3 7/8	.62326	.05193	4
5	1.475	1.748	6 1/8	5 7/8	5 1/4	5 3/16	5/8	3/4	3/8	1 13/32	.12	.635	1 1/2	4 15/16	.63151	.05262	5
6	2.116	2.494	8 9/16	8 1/4	7 3/8	7 1/4	3/4	1 1/8	1/2	2	.15	.760	1 3/4	7	.62565	.05213	6
7	2.750	3.270	11 5/8	11 1/4	10 1/8	10	1 1/8	1 3/8	3/4	2 5/8	.18	1.135	2 5/8	9 1/2	.62400	.05200	7

(b)

FIG. 13.12 (*a*) Cross-sectional view of lathe center with (*b*) dimensions.

When the setover tail stock method is used, the angle of the taper will vary with the length of the work. However, this problem may be overcome by using stock cut to identical lengths, with center holes drilled to the same depth. Under these conditions, the tail stock setover method has its advantages over other methods.

The compound rest is quite adequate for turning and boring short tapers and bevels, and for such work, it is probably the fastest method to set up and use. Merely set the compound rest swivel at the required angle and machine taper by rotating the compound rest feed screw by hand. In doing so, however, always make certain of the angle. Lathes vary in quality—and in accuracy—and some markings may be incorrectly calibrated or else not accurate enough for a particular requirement. Furthermore, it is sometimes difficult to read graduations with sufficient accuracy to set the compound rest swivel for an exact taper. Therefore, all angular or bevel turning should be tested with a gauge of some type before the cut is made.

A good example of taper turning with the compound rest is machining 60 degree centers for lathes. The compound rest can be swiveled and locked at any angle by means of the lockscrews. When the desired taper is expressed in degrees and minutes, the angle is simply transferred to the proper side of the 90 degree reading on the graduated base of the compound rest, which is exactly one-half the total included angle of taper. Therefore, in the case of a 60 degree center, the compound rest will be set at a 30 degree reading on the compound rest, as shown in Fig. 13.13.

When the taper is expressed in inches per foot, convert this figure into degrees and minutes. Find the tangent of the desired angle as follows:

Tangent of angle = taper per foot in inches / 24

Then consult tangent tables to obtain the equivalent of this tangent in degrees and minutes.

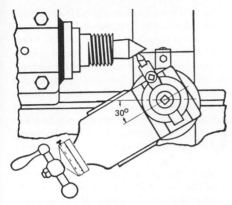

FIG. 13.13 Degree setting on compound rest for turning 60-degree centers.

Morse standard tapers are used for lathe and drill press spindles by most of the manufacturers of lathes and drill presses. The Myford Super 7 lathes, for example, have both head stock and tail stock spindles fitted for a No. 2 Morse taper which has a taper per foot of .59941 in. Using the previous equation, the tangent of the angle may be found as follows:

Tangent of angle = .59941 / 24 = .024975

From the appropriate table in this book, .024975 is the tangent of an angle of 1 degree, 26 minutes, or slightly less than 1½ degrees. Therefore, the correct compound rest setting is 1 degree, 26 minutes from the 90 reading.

Since the graduations on the compound rest cannot always be read exactly (minutes of an angle, for example) some sort of gage should be used when precision tapers are required. The Starrett center gages shown in Fig. 13.14 are good examples. These gages are available in three types: American National or U.S. 60 Standard, Whitworth or English 55 Standard, and 60 Metric Standard. Besides being useful for assuring 60 angles for turning lathe centers, they are also extremely handy for use in grinding and setting screw-cutting tools. These gages also have a very useful scale for finding the number of threads per inch or per mill by means of graduations in fourteenths, twentieths, twenty-fourths, and thirty-seconds of an inch on one type and in millimeters and half millimeters on another type. The American National center gage has in addition a table of double depths of threads for determining size of tap drills.

An attachment for these center gages is useful for holding the center gage firmly against the lathe arbor or face plate when setting both internal and external screw cutting tools. A slot containing a flat spring holds the gauge and a v groove in the lower side permits locating the attachment against the lathe arbor.

A universal bevel protractor is another useful tool. This protractor with vernier can be accurately read to 5 minutes or $\frac{1}{12}$ degree. The dial of the protractor is graduated both the right and

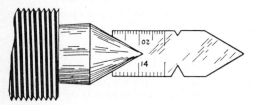

FIG. 13.14 The Starrett center gage is an excellent measuring device to obtain accurate angles when setting the compound rest to a given angle.

left of zero up to 90 degrees. The vernier scale is also graduated to the right and left of zero up to 60 minutes, each of the 12 vernier graduations representing 5 minutes. Since both the protractor dial and vernier scale have graduations in opposite directions from zero, any size angle can be measured, and it should be remembered that the vernier reading must be read in the same direction from zero as the protractor, either left or right.

Work that can be machined between centers can be taper-turned by setting over the tail stock as shown in Fig. 13.15. The amount of setover depends on the amount of taper per foot and also the overall length of the work. With the same amount of setover, pieces of different lengths will be machined with different tapers.

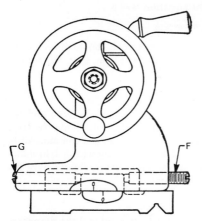

FIG. 13.15 Setting tail stock over for taper turning.

Setting the tailstock toward the operator (or tool post) results in a taper with the smaller diameter at the tail stock, while setting the tail stock in the opposite direction results in a smaller diameter at the head stock end of the work.

In determining the proper amount of tail stock setover, bear in mind that the tail stock center is set over one-half the total amount of the taper for the entire length of the work. To calculate the amount of setover, the following equations may be used:

When the taper per foot is given, such as inches per foot, the equation used is

Setover = taper per foot × length of taper in inches / 24

To demonstrate the use of this equation, assume that a piece of stock is exactly 1 ft long and is to be tapered ½ in per foot. Substituting these values in the equation, we have

Setover = .5 × 12/24 = .250 or ¼ in

To obtain the required taper, the tail stock should therefore be set over ¼ in. If the piece were only 10 in long, then the setover would be

Setover = .5 × 10 / 24 = .208 in

When the entire length of the piece is to be tapered and the diameters at both ends of the tapers are known, divide the large diameter, less the small diameter, by 2 to obtain the amount of setover. For example, a piece of stock 1 ft long with a 1-in diameter at one end and ½-in diameter at the other, should have the tail stock set over the following amount:

Setover = 1 − .5 / 2 = .250 or ¼ in

When a portion of the stock is to be tapered and the diameters at ends of the tapered portion are known, divide the total length of the stock by the length of the protion to be tapered and multiply this quotient by one-half the difference in diameters; the result is the amount of setover; that is,

Setover = total length of work / length to be tapered
$$\times \text{ large diameter } - \text{ small diameter } / 2$$

The method of adjusting the setover will vary slightly with the make of lathe, but in most cases all that is required is to loosen the clamping nut or lever and back off one of the setover screws which are located on each side of the tail stock body; then screw in the opposite setover screw until it is tight and reclamp the tail stock to the lathe bed. A zero mark is usually engraved at the end of the tail stock to serve as a rough guide to setover amounts, and to assist in returning the tail stock to its normal position for parallel turning.

To measure the setover of the tail stock center, place a scale having graduations on both edges between the two centers, as shown in Fig. 13.16. This will give an approximate measurement.

The best way to machine an accurate taper is to fit the taper to a standard gage as discussed previously. To test the taper, make a chalk mark along the entire length of the taper gage and then

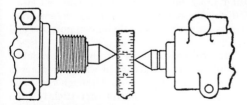

FIG. 13.16 Approximating the amount of tail-stock setover using a metal scale.

place the roughly tapered work into or onto the gage or the tapered piece the work is to fit; revolve the work carefully by hand. Then remove the work and the chalk mark will show where the taper is bearing so that adjustments can be made.

If the taper is a perfect fit, it will show along the entire length of the chalk mark. If the taper is not perfect, make the necessary adjustment, take another light chip and test again. Be sure the taper is correct before turning to the finished diameter.

The alignment of lathe centers should be checked at regular intervals, so as to maintain accuracy of long cuts, even though the tail stock might not have been purposely moved.

Taper Turning with Taper Attachment

A separate taper attachment is available for most lathes. It eliminates the necessity of setting over the tail stock, and if desired, may be set permanently for a standard taper, as it does not interfere with straight turning.

The taper-turning attachment shown in Fig. 13.17 has many advantages over the tail stock setover method. Lathe centers are never taken out of alignment; bearing surfaces of the lathe centers are not affected; duplicate tapers may be cut quickly on pieces of different length; and taper boring—impossible with tail stock setover—is handled quickly and easily. The index plate is graduated in degrees and taper per foot, simplifying computation and setting.

In general, there are two types of taper attachments available. One is known as the plain taper attachement while the other is known as the telescopic taper attachment. Although the exact method of attaching these to the lathe will vary slightly from manufacturer to manufacturer, all operate in basically the same manner. The attachment consists of a bracket attached to the

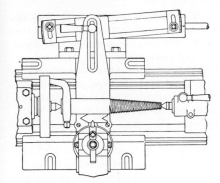

FIG. 13.17 Myford taper turning attachment.

back of the lathe carriage, a compound slide with clamp for locking it to the lathe bed and a connecting bar to connect the slide block of the taper attachment to the compound rest base of the lathe.

When the plain taper attachment is to be used it is necessary to disconnect the cross-feed screw by removing the bolt which locks the cross-feed nut to the compound rest base of the lathe. This leaves the compound rest base free to slide so that it may be controlled by the taper attachment.

The telescopic type is similar to the plain type except that the latter is equipped with a telescopic cross-feed screw which eliminates the necessity of disconnecting the cross-feed screw when the taper attachment is in use.

Most are assembled by first securing a bracket to the lathe bed at the desired position with machine screws. It should be positioned so that it will cover the portion of the work to be tapered since most will cover an area of about 6 in to about 1 ft on the smaller bench lathes. This bracket also contains the taper gage. Make sure that both the top and side surfaces of the guide bar are parallel with the ways of the lathe bed when the pointer is set at zero. This adjustment should be made with great care using an indicator to ensure accuracy. If the guide bar does not check parallel with the side of the lathe bed, set the guide bar parallel with the bed and move the graduated plate until the pointer and the zero mark are in line and fasten in place.

The drawbar is then secured to the cross-slide bed with the clamp provided and the drawbar opening is attached to the guide bar with the handle screw and washer, also provided. Before the taper attachment may be used, however, the cross-slide screw must be disengaged from the cross slide to enable the taper attachment to guide the crossbar at the required angle, which in turn will guide the tool holder and cutting bit. The chip guard is first removed from the cross slide by removing the two hex head screws (cap head screws) and then loosening the knock-off peg.

Pull the cross slide back to about the position it is to be used in and clamp the drawbar securely to the guide bar. The taper attachment is now ready for use.

14

Cutting Tools

Metal can be machined accurately and efficiently only with the correct type of tool—one having a sharp, well-supported cutting edge, ground for the application at hand. Furthermore, the tool must be positioned correctly and be set at the correct height.

LATHE CUTTING TOOLS

There are at least a dozen or more basic types of lathe cutting tools, and even more variations of these. Each has a specific application and each should be known. The basic lathe tools shown in Fig. 14.1 are those used mostly for metal cutting operations. These include:

Left-hand turning tool
Round-nose turning tool
Right-hand turning tool
Left-hand facing tool

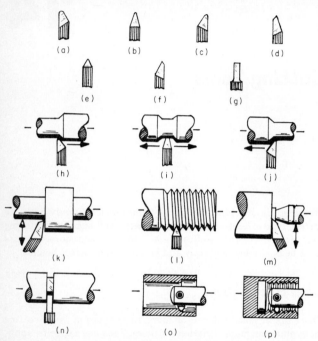

FIG. 14.1 Basic lathe cutting tools. (*a*) Left-hand turning tool, (*b*) round-nose turning tool, (*c*) left-hand facing tool, (*e*) threading tool, (*f*) right-hand facing tool, (*g*) cutoff tool, (*h*) left-hand turning tool, (*i*) round-nose turning tool, (*j*) right-hand turning tool, (*k*) left-hand facing tool, (*l*) threading tool, (*m*) right-hand facing tool, (*n*) cutoff tool, (*o*) boring tool, and (*p*) inside threading.

Threading tool
Right-hand facing tool
Cutoff tool
Boring tool
Inside-threading tool

GRINDING LATHE TOOLS

The cutting action of a lathe tool is shown in Fig. 14.2. The tool angles of clearance and rake are the most important for good performance. The clearance or angle of clearance of a lathe cutting tool allows the side of the tool to penetrate and feed along the work without rubbing or otherwise interfering below the cutting edge. In most cases, the side clearance angle should measure about 8 to 10 degrees; the front clearance angle should also measure the same.

The raking angles shown in Fig. 14.2 help improve the cutting action of the tool—in much the same way as tilting a knife blade helps in shaving wood. Tools designed for cutting soft metals, such as brass or copper, usually require little or no rake. However, clearance is always necessary regardless of the type of metal.

To grind a standard tool bit blank, as shown in Fig. 14.3, to the shape shown in Fig. 14.4, a definite grinding procedure should be followed. The tool in Fig. 14.4 is a standard right-hand facing tool, ground to feed toward the head stock of the lathe. A left-hand tool is fed in the opposite direction, or toward the tail stock. In grinding to shape, use the face of a relatively coarse, free-cutting grinding wheel (see Chap. 21), and use eye protection at all times. Then, follow these steps:

1. Grind the side clearance first as shown in Fig. 14.5, tilting the bit clockwise to remove more metal at the bottom; at the same

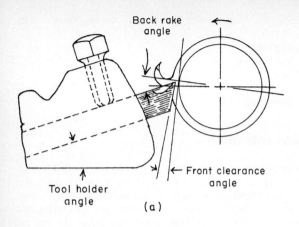

(a)

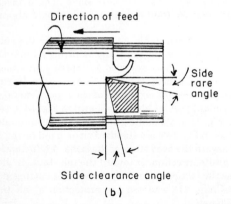

(b)

FIG. 14.2 (a) Back-rake and front-clearance angles in lathe cutting. (b) Side-rake and side-clearance angles in lathe cutting.

FIG. 14.3 Standard tool bit of high-speed steel.

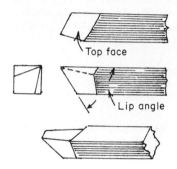

Top face

Lip angle

FIG. 14.4 Right-hand tool for mild steel.

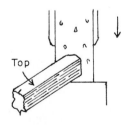

Top

FIG. 14.5 Grinding side clearance.

time, watch the shape of the top face. Use a ready-made bit as a guide as the grinding proceeds. The bit is gripped in both hands, with the top face up, resting the tool or hands on the tool rest.

2. Grind the front clearance, tilting the shank down as shown in Fig. 14.6, again observing the shape of the tool on top. The exact angles can be checked periodically with a shop protractor, or use a template or tool-grinding fixture. When grinding the front clearance, allow an extra 15 degrees for the tilt of the tool holder.

3. The side rake or slope of approximately 15 degrees is ground next as shown in Fig. 14.7. Use a medium grinding pressure to avoid undue heating, mostly to prevent tiny cracks if the tool is then quenched in water.

4. Round off the point slightly, by swinging the tool in an arc sideways as shown in Fig. 14.8. A radius of about ⅓₂ in should

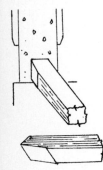

FIG. 14.6 Grinding the front clearance.

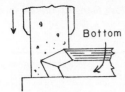

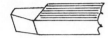

FIG. 14.7 Grinding the side rake of about 15 degrees.

be sufficient for a trial. This arc must extend all the way down the front of the bit or else the front clearance will be spoiled.

When completed, check all angles, and if any are found to be correct, hone the bit either by hand or with a polishing wheel,

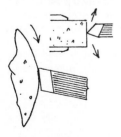

FIG. 14.8 Rounding off the tool point.

without rounding off the top of the bit. Also note the following table of rake angles for various materials.

Material being worked	Side rake, degrees	Back rack, degrees
Mild steel	15	15
Cast iron	13	6
High-carbon steel	11	6
Soft Brass	0	0
Hard bronze	4	3
Aluminum	16	30
Copper	23	13

MILLING CUTTERS

In a milling machine, the cutter rotates and the work is fed against it; just the opposite from the metal-cutting lathe. The milling cutter is made in almost an unlimited variety of shapes and sizes for milling practically any shape imaginable. Some types of milling cutters are shown in Fig. 14.9.

Milling cutters are made from tool steel, heat-treated to the proper hardness, and then finished to the specific dimensions by

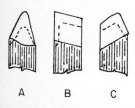

FIG. 14.9 Some types of milling cutters. (*a*) Round-nose tool, (*b*) right-hand corner tool, (*c*) left-hand turning tool.

grinding. The teeth of milling cutters are given a slight clearance back from the cutting edge of from 4 to 7 degrees for peripheral teeth (teeth on the circumference of a cutter), and 2 or 3 degrees for side teeth.

REAMERS

Reamers are used to enlarge a drilled hole to a specific diameter and with a reasonable degree of accuracy. The fluted reamer is one having numerous flutes on the circumference of the cutting portion of the tool. The cutting takes place along the long taper on a hand reamer and along the 45-degree bevel on a machine reamer. The number of flutes on the surface of a reamer varies with the diameter.

Hand reamers are usually tapered toward the point for a short distance to give a lead into the drilled hole to facilitate the hand-reaming operation. Reamers are made from tool steel and heat-treated to the proper hardness before final grinding to size. The grinding process is done with extreme accuracy—usually in a grinding machine specially designed for the purpose.

Regrinding reamers is not recommended for the average shop without the proper equipment. Rather, they may be sharpened by honing. The crests of the teeth should be honed with an Arkansas stone. Resharpening in this manner prolongs the useful life of the reamer with little reduction in its effective diameter. If the reamer in question has become seriously blunted, it is best to have the tool reground by an expert toolmaker.

15

Taps and Dies

Taps are tools used to form internal threads while dies are used to form outside threads.

Machinists will encounter two types of holes: the through-hole and the blind hole (see Fig. 15.1). The through-hole is one that has been drilled entirely through the material—allowing the tap to pass completely through. Blind holes are shallow holes drilled in material, necessitating the tap to be stopped before hitting bottom, otherwise tap breakage will occur.

TAPS

Taps are screw-like tools with two, three, or four flutes. The cutting threads are on one end followed by a round body with four flats on the end of the shank. These four flats engage in the tap chuck to assure a positive driving action.

Some of the more common types of taps are hand, machine,

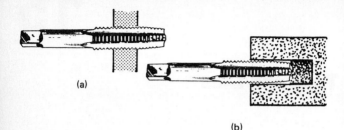

FIG. 15.1 (*a*) Through-hole and (*b*) blind-hold taos.

pipe, stay bolt, boiler, and pulley taps. Taps are usually made in sets of three:

Taper tap (no. 1 tap).   Tapered to facilitate its entrance into the hole, and so the metal will be removed gradually to form the thread.

Plug tap (no. 2 tap).   After the thread is started, it may be carried through with the plug tap. If the hole is a blind hole, it may be necessary to start it with the plug tap.

Bottom tap (no. 3 tap).   After running the plug tap in a blind hole, clean the hole out and finish the thread to the bottom of the hole with the bottom tap.

The above is the correct use for the taps described. However, in some applications involving through-holes, the taper tap is often passed up, going directly to the plug tap or bottom tap, making one pass do for the complete tapping operation.

In general, two styles of taps are currently in use: the standard hand tap and the spiral-fluted machine-screw tap. The former is used in tapping a hole by hand, usually started with a taper tap so very little force is needed to urge the tap through. If care is taken to start the tap straight—held vertically and at right angles to work surface—it will draw itself through as the thread is cut. Once the taper tap has been used, it is followed with the plug tap.

In machine tapping, a tap holder must be used. Spiral flutes are used to help draw the chips out of the hole. Spiral taps are most frequently used for blind holes, especially in stringy material such as aluminum, magnesium, brass, copper, and die-cast metals. In addition there is the gun tap, used on through holes, that shoots chips out ahead. Flutes are not needed for chip disposal in the gun tap, so the flutes may be made shallower, thereby increasing the strength of the tap. Because of machine feeding, the plug tap is used mostly for machine tapping.

Cast iron and brass are normally tapped dry but oil should be used freely on steel and wrought iron.

In drilling a hole to accommodate any tap, it is important that the hole be drilled to the proper size, or the thread may be cut too deep or not deep enough. The average hole is tapped so that approximately 75 percent of the screw thread is engaged with the hole thread. There are certain applications where it might be desirable to have more or less thread engaged. This is accomplished by using a larger or smaller drill bit. Table 15.1 shows taps and corresponding drill sizes for Unified and American screw threads.

DIES

Unlike a tap that is used to cut internal threads, a die is used to cut external threads on bolts, rods, and shafts. The type of thread-cutting die most often used is called an adjustable or split

TABLE 15.1 American Standard and Unified Form Threads

Thread size	Pitch series	Tap drill size	Decimal equivalent of tap drill	Percent of thread (approx.)
0–80	NR	56	.0465	83
		3/64	.0469	81
1–64	NC	54	.0550	89
		53	.0595	67
1–72	NR	53	.0595	75
		1/16	.0625	58
2–56	NC	51	.0670	82
		50	.0700	69
		49	.0730	56
2–64	NF	50	.0700	79
		49	.0730	64
3–48	NC	48	.0760	85
		5/64	.0781	77
		47	0.785	76
		46	.0810	67
		45	.0820	63
3–56	NF	46	.0810	78
		45	.0820	73
		44	.0860	56
4–40	NC	44	.0860	80
		43	.0890	71
		42	.0935	57
		3/32	.0938	56
4–48	NF	43	.0890	85
		42	.0935	68
		3/32	.0938	67
		41	.0960	59
5–40	NC	40	.0980	83
		39	.0995	79
		38	.1015	72
		37	.1040	65
5–44	NF	38	.1015	80
		37	.1040	71
		36	.1065	63
6–32	NC	37	.1040	84
		36	.1065	78

Thread size	Pitch series	Tap drill size	Decimal equivalent of tap drill	Percent of thread (approx.)
		14	.1820	73
		13	.1850	67
¼–20	UNC	9	.1960	83
		8	.1990	79
		7	.2010	75
		¹³⁄₆₄	.2031	72
		6	.2040	71
		5	.2055	69
¼–28	UNF	3	.2130	80
		⁷⁄₃₂	.2188	67
⁵⁄₁₆–18	UNC	F	.2570	77
		G	.2610	71
⁵⁄₁₆–24	UNF	H	.2660	86
		I	.2720	75
		J	.2770	66
⅜–16	UNC	⁵⁄₁₆	.3125	77
		O	.3160	73
⅜–24	UNF	Q	.3320	79
		R	.3390	67
⁷⁄₁₆–14	UNC	T	.3580	86
		²³⁄₆₄	.3594	84
⁷⁄₁₆–20	UNF	W	.3860	79
		²⁵⁄₆₄	.3906	72
½–13	UNC	²⁷⁄₆₄	.4219	78
½–20	UNF	²⁹⁄₆₄	.4531	72
⁹⁄₁₆–12	UNC	¹⁵⁄₃₂	.4688	87
		³¹⁄₆₄	.4844	72
⁹⁄₁₆–18	UNF	½	.5000	87
		0.5062	.5062	78
⅝–11	UNC	¹⁷⁄₃₂	.5312	79
⅝–18	UNF	⁹⁄₁₆	.5625	87
		0.5687	.5687	78
¾–10	UNC	⁴¹⁄₆₄	.6406	84
		²¹⁄₃₂	.6562	72
¾–16	UNF	¹¹⁄₁₆	.6875	77
⅞–9	UNC	⁴⁹⁄₆₄	.7656	76

TABLE 15.1   *(Continued)*

Thread size	Pitch series	Tap drill size	Decimal equivalent of tap drill	Percent of thread (approx.)
		$\frac{7}{64}$	.1094	70
		35	.1100	69
		34	.1110	67
		33	.1130	62
6–40	NF	34	.1110	83
		33	.1130	77
		32	.1160	68
8–32	NC	29	.1360	69
8–36	NF	29	.1360	78
		28	.1405	65
		$\frac{9}{64}$	.1406	65
10–24	NC	27	.1440	85
		26	.1470	79
		25	.1495	75
		24	.1520	70
		23	.1540	66
10–32	NR	$\frac{5}{32}$	.1562	83
		22	.1570	81
		21	.1590	76
		20	.1610	71
12–24	NC	$\frac{11}{64}$	.1719	82
		17	.1730	79
		16	.1770	72
		15	.1800	67
12–28	NR	16	.1770	84
		15	.1800	78

Thread size	Pitch series	Tap drill size	Decimal equivalent of tap drill	Percent of thread (approx.)
7/8–14	UNF	51/64	.7969	84
		0.8024	.8024	78
		13/16	.8125	67
1–8	UNC	55/64	.8594	87
		7/8	.8750	77
1–12	UNF	29/33	.9062	87
		59/64	.9219	72
1 1/8–7	UNC	31/32	.9688	84
		63/64	.9844	76
1 1/8–12	UNF	1 1/32	1.0312	87
		1 3/64	1.0469	72
1 1/4–7	UNC	1 3/32	1.0938	84
1 1/4–12	UNF	1 5/32	1.1562	87
		1 11/64	1.1719	72
1 3/8–6	UNC	1 3/16	1.1875	87
		1 13/64	1.2031	79
		1 7/32	1.2188	72
1 3/8–12	UNF	1 9/32	1.2812	87
		1 19/32	1.2969	72
1 1/2–6	UNC	1 5/16	1.3125	87
		1 21/64	1.3281	79
1 1/2–12	UNF	1 13/32	1.4062	87
		1 27/64	1.4219	72
1 3/4–5	UNC	1 17/32	1.5312	84
		1 35/64	1.5469	78
2–4 1/2	UNC	1 25/32	1.7812	76

die. This type of die can be adjsuted to cut a thread to the kind of fit desired, such as loose, snug, or tight.

The two-piece die is made in two parts that fit into a die stock especially designed for this type of die. A solid square die is sometimes used as are the precision die stock. The bolt, rod, or other object to be threaded is first given a short chamfer or angle filed on the end to help start the die easily. Of course, most dies are chamfered back a distance of three or four threads to facilitate starting a thread. Always start the chamfered side of the die on the work first. Then, if it is desired to thread as close to the shoulder as possible, the die is turned over where only the one thread is chamfered.

To cut an external thread, grind or file a chamfer or bevel on the end of the rod or bolt to help the die start more easily. Secure this rod in a bench vise, preferably in a vertical position. Then secure the die in the die stock with the stamped size on the top. The threads on the opposite side are tapered to help start the die. If the die stock has a guide, adjust it for a free fit on the rod. Place the die over the end of the rod, apply some pressure, and turn it to get the first threads started. Apply cutting fluid with a brush or oil can so that oil gets down into the threads, then check the alignment of the die before continuing.

After threading for two or three complete turns, back the die off a half turn or so to break the chips. Once the thread is cut to the correct length, back the die off by turning the die holder counterclockwise.

Check the thread with a gage, a nut, or the part it is to fit into. If the thread is too tight, close the die slightly by backing up the setscrew in the die slot and tightening the opposite setscrews in the die stock. Then run the die over the thread again. Performing this operation once or twice should size the thread correctly.

METRIC THREADS

Metric taps and dies are used when metric threads are required. The International Standards Organization (ISO) metric threads are specified in two series: coarse and fine. However, for all practical purposes, only the coarse series is used in threading and assembling. The fine thread series is used only for precision work such as the manufacture of microscopes and other fine instruments.

In the coarse sizes, the 12 common sizes range in diameter from 2 to 24 mm, with a pitch range from 0.4 to 3.0 mm. Metric gages are available to quickly determine the size.

TAP EXTRACTORS

Regardless of the care used in tapping drilled holes, there is always the possibility of breaking a tap. If the break is down into the hole, removing the tap is often very difficult since taps are heat-treated to a very hard surface, and most cannot be drilled with ordinary bits. If the break is above the hole, the job is relatively easy since it can usually be gripped with a pair of pliers and backed out of the hole. Breaks below the surface can only be removed with a tap extractor.

To use a tap extractor, thoroughly remove all chips of the broken tap before inserting the extractor fingers into the flutes of the broken tap—pushing them gently but firmly into position. Next push the holder down until it touches the broken tap. Slide the sleeve down until it touches the work. Both of these steps are extremely important. Apply a tap wrench to the square end of the holder. Twist forward and backward a few times to loosen, then back out the broken tap.

16

Drilling Operations

TWIST DRILLS

There are many kinds of drills ranging from those for general purposes to specialized types for specific materials or applications. Various lengths and numbers of flutes are available, with straight or angled flutes, with regular or special points, and with straight or rapered shanks (see Fig. 16.1).

Flat drill. This type is among the simplest of drills, and is often made in the shop. It can be made by flattening, hardening and sharpening a good grade of steel and will then bore a satisfactory hole of shallow depths where precision is of no great concern.

Twisted drill. The twisted drill is essentially the same as a flat drill, but twisted to remove chips. It is made from flat stock, usually rolled thinner in the middle than at the edges, and then twisted. Some advantages of this drill over the twist drill are that

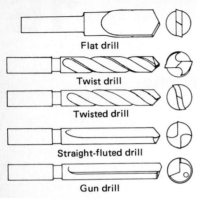

Flat drill

Twist drill

Twisted drill

Straight-fluted drill

Gun drill

FIG. 16.1 Typical drill types in common use.

it is more economical to make, has ample chip clearance and can be more readily hardened and tempered.

Twist drill. This is the most commonly used drill and is made from carbon or alloy steel, although these alloy steel bits are normally referred to as high-speed-steel bits. They are made from round stock with flutes cut by a milling cutter or formed by rolling or forging. They have two, three, or four cutting edges. The most common type is the two-lipped twist drill.

The twist drill is used for practically all general work, but like most useful tools, its accuracy depends upon the care with which it is prepared and used. To establish a cutting edge and allow for feed, the drill must be ground for lip clearance, and the lips must be alike.

Elimination of drill breakage depends a lot on the correct point grinding of the drill. When the drill is worn down by use, it must be reground or breakage will result. Not only must the drill be reground in time, but it must be properly ground to assure maximum drilling life. It has been estimated that 90 percent of all drill breakage is caused by incorrect regrinding, and for this reason, too much emphasis cannot be placed on the importance of this operation.

Point grinding by hand requires great skill and care on the part of the operator, and whenever skilled help is not available, the use of a point-grinding jig is recommended. If the drill is accidentally allowed to get too hot during grinding, it should never be cooled off in water, but should be allowed to cool of its own accord. A sudden cooling is almost sure to result in breakage.

When repointing a drill, four things must be carefully considered: (1) lip clearance, (2) point angle, (3) cutting edges, and (4) point thinning.

The cutting edges, or lips, should make a uniform and exact angle with the axis of the drill, and that angle should be 59 degrees, except under unusual conditions when larger angles are sometimes used. Next, the cutting edges must be the same length and the clearance angle back from the cutting lips may be from 12 to 15 degrees. Best clearance for general work is 12 degrees—for soft material, 15 degrees. If sharpening is accurate, the drill will run true.

A properly sharpened drill bit will only allow so many thousandths of an inch material to be removed per each revolution of the bit. Only moderate pressure is required to make the bit function properly. Excess pressure only adds to operator fatigue and can damage the drill bit, as the excess pressure does not make the bit drill faster or remove more material.

To eliminate spinout of a bit on smooth surface, it is recommended that the spot first be center-punched to assure proper positioning of the drill point.

When drilling large diameter holes, it is best to first drill a smaller (pilot) hole and then follow with the larger bit.

Twist drills under ½ in are almost always made with straight shanks, while shanks on larger bits may be either straight or tapered. Tapered shanks usually have the Morse taper.

Figure 16.2 shows the recommended drill points for drilling various materials.

The speeds and feeds listed in Table 16.1 apply to average working conditions and materials. They are recommended with a due regard to conserving drills and avoiding excessive machine tool wear.

Under many conditions, these speeds and feeds may be considerably increased, while under others they must be reduced. In order to secure best results, both the speed and feed should be increased or decreased in proper proportion. The liberal use of cooling compound (Table 16.2) will increase the life of tools.

RADIAL DRILLING

The radial drill is a frequently used tool for mass production of duplicate work. When used with a precision jig, a radial drill is capable of drilling and tapping a succession of holes in a metal workpiece with one clamping of the work. One chief advantage of the radial drill over the conventional drill press is its horizontal arm, which can be swung radially about its upright standard and carries a spindle that can be moved back and forth. It is this movement or arm and spindle that permits drilling successive holes without shifting the work.

For precision, and to save layout time, jigs are used in conjunction with the radial drill. Jigs are normally made by an expert toolmaker, who locates, drills, and sometimes grinds the holes that guide the bit into the work. These holes are then made to fit a hardened steel bushing, which in turn fits the bit to be used and keeps it from marring the jig. The following is a typical operating procedure for the radial drill:

1. Clean the drill table of all dirt, chips, and burrs before proceeding. This will help keep the holes straight and will prevent breaking the bit during the drilling.

2. Select a jig and measure the holes to make certain it suits the job at hand.

3. Place the workpiece on the drill table, being careful to avoid damage. Place the jig in position and secure both the work and jig to the table with hold-down clamps and blocks.

4. Match lineup markings—that is, match lines on the jig to lines on the workpiece.

5. Once the work and jig have been set up, choose a drill chuck and taper and secure them in place.

6. Select the correct-size tap drill and set it into a friction chuck, which puts a limited torque on the bit while it is running in the work. This prevents breakage should it jam.

7. When drilling blind holes, set the spindle stop on the drill head to the exact depth to arrest the feed automatically at that depth.

8. Refer to appropriate chart for the correct rpm for the drill and type of metal used and the correct feed rate. Controls or gears on the drill are then set for the correct rate for each operation.

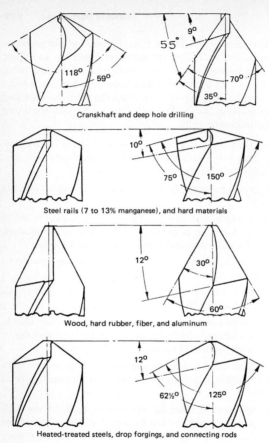

Cranskhaft and deep hole drilling

Steel rails (7 to 13% manganese), and hard materials

Wood, hard rubber, fiber, and aluminum

Heated-treated steels, drop forgings, and connecting rods

FIG. 16.2 Drill points recommended for various applications.

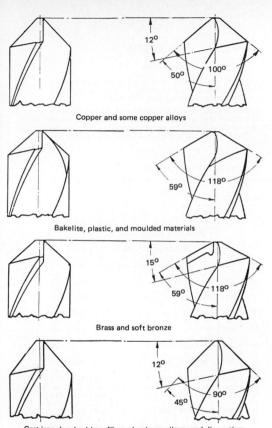

Copper and some copper alloys

Bakelite, plastic, and moulded materials

Brass and soft bronze

Cast iron, hard rubber, fiber, aluminum alloys, and die castings

FIG. 16.2 *(Continued).*

TABLE 16.1

SPEEDS AND FEEDS

For High Speed Drills

The speeds and feeds in the following table apply to average working conditions and materials. They are recommended with a due regard to conserving drills and avoiding excessive machine tool wear.

Under many conditions, these speeds and feeds may be considerably increased, while under others they must be decreased. In order to secure best results, both the speed and feed should be increased or decreased in proper proportion. The liberal use of cooling compound will increase the life of tools.

KEEP DRILLS SHARP

MATERIAL	Speed in Feet per Minute*	DIAMETER OF DRILLS								
		$\frac{1}{16}$ to $\frac{1}{8}$	$\frac{5}{32}$ to $\frac{1}{4}$	$\frac{9}{32}$ to $\frac{3}{8}$	$\frac{13}{32}$ to $\frac{1}{2}$	$\frac{17}{32}$ to $\frac{3}{4}$	$\frac{25}{32}$ to 1	$1\frac{1}{32}$ to $1\frac{1}{4}$	$1\frac{9}{32}$ to $1\frac{1}{2}$	$1\frac{1}{2}$ and larger
		FEED PER REVOLUTION INCHES								
Aluminum	250 to 300	.002 to .003	.003 to .005	.005 to .006	.006 to .008	.008 to .010	.010 to .013	.013 to .015	.015 to .016	.016 to .018
Cast Iron, Malleable Iron	75 to 110	.002 to .004	.004 to .006	.006 to .009	.009 to .012	.012 to .016	.014 to .020	.018 to .025	.022 to .028	.022 to .030
Brass or Bronze	150 to 200	.002 to .004	.004 to .007	.007 to .010	.010 to .014	.014 to .018	.016 to .022	.020 to .026	.026 to .030	.030 to .040
Monel Metal, Stainless, Molybdenum, Drop Forged Alloys, Tool Steel, Annealed	60 to 75	.002 to .003	.003 to .004	.004 to .006	.006 to .009	.008 to .012	.010 to .014	.012 to .016	.014 to .020	.016 to .026
Drop Forgings Alloy Steel Heat Treated	45 to 60	.002 to .003	.003 to .004	.004 to .005	.005 to .006	.007 to .010	.009 to .012	.010 to .014	.012 to .016	.014 to .022
3½% Nickel Steel, Steel Castings	60 to 75	.002 to .003	.003 to .005	.004 to .006	.006 to .010	.010 to .014	.014 to .016	.016 to .022	.018 to .025	.020 to .028
Mild Steel .2 to .3 Carbon	70 to 95	.002 to .003	.003 to .005	.005 to .007	.006 to .010	.010 to .014	.014 to .016	.016 to .022	.018 to .025	.020 to .028

Table of Cutting Speeds
Feet per Minute

Diam. Ins.	15	20	25	30	35	40	45	50	55	60	70	75
	Revolutions Per Minute											
1/32	1835	2445	3060	3668	4280	4890	5500	6114	6726	7336	8560	9172
1/16	917	1222	1528	1834	2140	2445	2751	3057	3363	3668	4280	4586
1/8	458	611	764	917	1070	1222	1375	1528	1681	1834	2139	2292
3/16	306	407	509	611	713	815	917	1019	1121	1222	1426	1529
1/4	229	306	382	458	535	611	688	764	851	917	1070	1147
5/16	183	244	306	367	428	489	550	611	672	733	856	917
3/8	153	204	255	306	357	408	458	509	560	611	713	764
7/16	131	175	218	262	306	349	393	437	481	524	611	656
1/2	115	153	191	229	268	306	344	382	420	459	535	573
9/16	103	137	172	204	238	272	306	340	373	407	475	509
5/8	91.7	122	153	184	214	245	276	306	337	367	428	459
11/16	83.5	112	140	167	194	222	249	273	300	333	389	416
3/4	76.4	102	127	153	178	203	229	254	279	306	357	381
13/16	71.0	95.0	118	142	166	190	213	237	261	284	332	356
7/8	65.5	87.3	109	131	153	175	196	219	241	262	306	329
15/16	61.4	82.0	102	122	142	163	183	204	224	244	285	305
1	57.3	76.4	95.5	115	134	153	172	191	210	229	267	287
1 1/8	50.9	67.9	84.9	102	119	136	153	170	187	204	238	255
1 1/4	45.8	61.1	76.4	91.8	107	123	137	153	168	183	214	230
1 3/8	41.7	55.6	69.5	83.3	97.2	111	125	139	153	167	195	208
1 1/2	38.2	50.9	63.7	76.3	89.2	102	115	127	140	153	178	191
1 5/8	35.3	47.0	58.8	70.5	82.2	93.9	106	117	129	141	165	176
1 3/4	32.7	43.7	54.6	65.5	76.4	87.3	98.2	109	120	131	153	164
1 7/8	30.6	40.7	50.9	61.1	71.3	81.5	91.9	102	112	122	143	153
2	28.7	38.2	47.7	57.3	66.9	76.4	86.0	95.5	105	115	134	143
2 1/8	27.1	36.0	44.0	54.0	63.0	72.0	81.0	90.0	99.0	108	126	135
2 1/4	25.5	34.0	42.4	51.0	59.4	68.0	76.2	85.5	93.5	102	119	128
2 3/8	24.2	32.3	40.3	48.3	56.4	64.4	72.5	80.5	88.6	96.6	113	121
2 1/2	22.9	30.6	38.2	45.8	53.5	61.2	68.8	76.3	81.2	91.7	107	114
2 5/8	21.8	29.2	36.4	43.5	50.8	58.0	65.3	72.5	79.8	87.0	102	109
2 3/4	20.8	27.8	34.7	41.7	48.6	55.6	62.5	69.6	76.5	83.4	97.2	104
2 7/8	20.0	26.6	33.2	39.6	46.2	52.8	59.4	66.0	72.6	79.2	92.4	99.0
3	19.1	25.5	31.8	38.2	44.6	51.0	57.3	63.7	69.9	76.4	89.1	95.3
3 1/4	17.6	23.5	29.4	35.1	41.0	46.8	52.7	58.5	64.4	70.2	81.9	87.8
3 1/2	16.4	21.8	27.3	32.7	38.2	43.6	49.1	54.5	60.0	65.5	76.4	81.8
3 3/4	15.3	20.4	25.5	30.6	35.7	40.8	45.9	51.0	56.1	61.2	71.4	76.5
4	14.3	19.1	23.9	28.7	33.4	38.2	43.0	47.8	52.6	57.3	66.9	71.7
4 1/2	12.7	17.0	21.2	25.5	29.7	34.0	38.2	42.4	46.7	50.9	59.4	63.6
5	11.5	15.3	19.1	22.9	26.7	30.6	34.4	38.2	42.0	45.8	53.5	57.3
5 1/2	10.4	13.9	17.4	20.8	24.3	27.8	31.3	34.7	38.2	41.7	48.6	52.1
6	9.5	12.7	15.9	19.1	22.3	25.5	28.6	31.8	35.0	38.2	44.6	47.8
6 1/2	8.8	11.8	14.7	17.6	20.6	23.5	26.4	29.4	32.3	35.3	41.1	44.1
7	8.2	10.9	13.6	16.4	19.1	21.8	24.5	27.3	30.0	32.7	38.2	40.9
7 1/2	7.6	10.2	12.7	15.3	17.8	20.4	22.9	25.5	28.0	30.6	35.7	38.2
8	7.2	9.5	11.9	14.3	16.7	19.1	21.5	23.9	26.3	28.7	33.4	35.8
8 1/2	6.7	9.0	11.2	13.5	15.7	18.0	20.2	22.5	24.7	27.0	31.5	33.7
9	6.4	8.5	10.6	12.7	14.9	17.0	19.1	21.2	23.3	25.5	29.7	31.8
9 1/2	6.0	8.0	10.1	12.1	14.1	16.1	18.1	20.1	22.1	24.1	28.2	30.2
10	5.7	7.6	9.5	11.5	13.4	15.3	17.2	19.1	21.0	22.9	26.7	28.7
11	5.2	6.9	8.7	10.4	12.2	13.9	15.6	17.4	19.1	20.8	24.3	26.0
12	4.8	6.4	8.0	9.5	11.1	12.7	14.3	15.9	17.5	19.1	22.3	23.9

SOURCE: Chicago Latrober.

TABLE 16.2 Oils and Compounds for Machining Operations

Material	Drilling	Reaming	Turning	Milling	Threading
Aluminum	Kerosene, kerosene and lard oil, soluble oil	Kerosene, soluble oil, mineral oil	Soluble oil	Soluble oil, lard and mineral oils, dry	Soluble oil, kerosene and lard oil
Brass	Dry, soluble oil, kerosene and lard oil	Dry, soluble oil	Soluble oil	Dry, soluble oil	Lard oil, soluble oil
Bronze	Soluble oil, lard oil, mineral oil, dry	Soluble oil, lard oil, mineral oil, dry	Soluble oil	Soluble oil, lard oil, mineral oil, dry	Lard oil, soluble oil
Cast iron	Dry, air jet, soluble oil	Dry, soluble oil, lard mineral oil	Dry, soluble oil	Dry, soluble oil	Oil, mineral lard oil
Cast steel	Soluble oil, mineral lard oil, sulfurized oil	Soluble oil, mineral lard oil, lard oil	Soluble oil	Soluble oil, mineral lard oil	Mineral lard oil
Copper	Soluble oil, dry, mineral lard oil, kerosene	Soluble oil, lard oil	Soluble oil	Soluble oil, dry	Soluble oil, lard oil

SOURCE: Chicago Latrobe.

9. With a hardened bushing on the bit and resting on the jig, move the drill head about until the bushing drops into a hole in the jig. Lock the arm and leg of the radial drill in place and begin drilling with hand feed.

10. As soon as the bit has made a good start, raise the spindle by hand and remove the bushing. Then the drilling proceeds with automatic feed.

11. When all holes have been completed, loosen the clamps, remove the jig, and clean the table. Put a plate of the same thickness as the jig under the work and use the same blocks to clamp it for tapping.

Depending upon the finish required on the object being drilled in the radial drill, further operations may include filing, grinding, and the like. Also, many of the operations described above are now handled automatically by computer-controlled drills, especially in the automotive and firearms industries.

DRILLING IN THE LATHE

Many drilling and reaming jobs can be performed with the conventional metal-turning lathe, and sometimes more accurately and more quickly than by most other methods. A drill pad is normally used for standard drillling applications. The pad is placed in the tail stock spindle to support the work. The bit is then inserted in a crull chuck screwed onto the head stock. Most drilling work on the lathe, however, is done with the work mounted in the lathe chuck or clamped to the faceplate of the lathe.

When the above method is used, it is important that the drill be started so that it will run true and the hole will be drilled concentric with the outside diameter of the work.

One method for starting the drill point true is as follows. Position the butt end of a lathe tool holder so it is just touching the side of the drill. This will prevent the drill from bending and cause it to start approximately true in the center of the work.

17

Shapers and Planers

Metal shapers and planers are metal-removing machines that are useful in any machine shop. The cutting tool in the shaper is moved in a horizontal plane by a ram having a reciprocating motion, cutting only on the forward stroke much like a hacksaw blade. The work being cut is held in a vise or otherwise clamped to the box-like table which moves either vertically or horizontally to obtain the desired shape of the work piece. The two types in common use are the "crank" and the "geared" shaper.

The planer is a machine tool that is used to machine flat or plane surfaces on work that is fastened to a reciprocating table. The surfaces can be horizontal, vertical, or angular. In some cases, the planer may also be used to form irregular or curved surfaces.

While these machines are very similar, the main difference is that in the shaper the tool moves while the work is stationary. In the planer, the work vibrates while the tool remains stationary, except for transverse movements in both.

Basic parts of the shaper are the base, crossrail, saddle, table, ram, and bull gear as shown in Fig. 17.1. The base supports the machine and all working parts while the crossrail contains the table elevating and traversing mechanism and is attached to the base by means of gib plates and bolts. Good operation on the shaper involves selecting the proper cutting tool, holding the work securely, adjusting the work accurately, adjusting the stroke, and selecting the proper feed and cutting speed.

The planer is used for practically the same purpose as the shaper except the planer can handle much larger work and take heavier cuts. The most common type of planer is the double-column machine where the columns support the crossrail and house the elevating screws and controls. Another type is the open-side

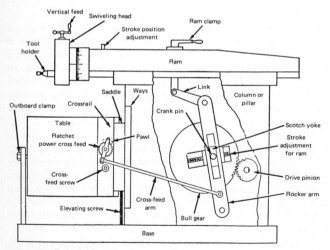

FIG. 17.1 Principal parts of a metal shaper.

planer which has only one column or housing to support the cross-rail and toolhead. One big advantage on this latter type is that workpieces of irregular shape can be handled with the workpiece extending outward over the side of the table.

Cutting tools used on both are similar to those used on the lathe, the main difference being that lathe tools tend to spring away from the work while the planer tends to dig in. Of course, the lathe is used to machine round objects while the planer is used for flat work.

SHAPER OPERATION

The first order of business when working with a shaper is to adequately hold the work. For small workpieces, either a vise or chuck can be used. When work is bolted directly to the table, it is important to place the T bolts properly or else the tool will merely push the work rather than cutting it. However, the bolts should not be tightened more than necessary to avoid distorting or springing the work. Since the cutting tool on a shaper does not tend to elevate the workpiece, extreme tightening of the bolts is not necessary.

Other items used to hold work on the shaper table are stop pins and toe dogs, stop pins and a table strip, various shaped braces, and special fixtures.

After the workpiece has been mounted and aligned on the shaper table, the table will be elevated until the workpiece clears the ram a distance of approximately 1 in. The cutting tool should also be clamped firmly at this time, and it should be made certain that the tool clears the work.

The ram has to be adjusted to provide the proper length of stroke and to provide the proper position of travel over the workpiece. The stroke should be long enough for the cutting tool to clear the work by not less than ¼ in on the forward stroke and by ½ in on the return stroke; this enables the automatic feed to function before the cutting tool comes in contact with the work.

Movement of the cutting tool and the tool slide is controlled by the vertical feed handle at the top of the head. During operation, the operator should decide on the number of cuts necessary to remove the desired amount of metal. In most cases, one or two roughing cuts and the same number of finishing cuts are all that will be necessary.

In selecting the feed for the shaper, the type of metal will have a great bearing on the feed selected. Of course, the capacity of the shaper must also be considered, as heavier feeds can be used on a large heavy-duty machine than on a small and relatively lightly constructed shaper. In general, the size of a shaper is determined by the size of the largest cube that can be machined on it. Also, machinists can affect the efficiency of a shaper by their selection of the cutting tool, method of holding the work, adjustment of the work, adjustment of the stroke, and selection of the proper feed and cutting speed.

Two types of planers are in common use, the double-column planer and the open-side planer. In either case, their capacity is judged by the width, height, and length of the largest workpiece that can be machined.

Cutting tools used on the planer are similar to those used on a conventional metal-turning lathe. However, the planer is used to machine flat surfaces, while circular surfaces are machined in lathe operations.

The method used to hold the work on the planer table is very important. Some clamping devices in common use are bolts, studs, washers, shims, nuts, step blocks, toe dogs, strap clamps, c clamps, and stops.

Another related machine is the slotting machine, which may be called a vertical shaper. With the slotting machine, the ram moves in a vertical direction rather than horizontally as on the shaper. In use, the slotter can be used for a variety of work other than slotting. Regular and irregular surfaces, both internal and external, can be readily machined on the slotter. It is especially adaptable for handling heavy pieces of work that cannot be handled easily on other machines.

A typical set of slotter cutting tools include the roughing, finishing, keyway, and scriber tools. This last tool is used mainly for cylindrical work setups. In addition, several special tools may be ground for specialized work.

CONTROL OF MACHINE TOOLS

Most modern manufacturing plants are utilizing electronic control of machine tools for as many applications as possible. In general, a recording is made of the various movements required of a particular tool, for a specific series of operations required to produce a machine part. The recording, which is translated from a drawing of the part, is recorded on magnetic tape or programmed into a computer system.

Four basic steps are normally involved when implementing an automatic machine tool system in industry:

1. Development of the working drawings
2. Planning
3. Preparation of a computer program or control tape
4. Coordinating the program to operate the machine

18

Milling

The milling machine is a very versatile power-driven machine tool that performs a great variety of machining operations with speed and accuracy. It is particularly useful when a large number of interchangeable workpieces must be cut to exact dimensions, and its adaptability is such that it can handle both large and small parts of intricate design.

The two main elements of the milling machine are a movable table and a revolving cutter. The table carries the work to a rotating cutter, the latter remaining in a fixed position while the work moves to form the shape or design. However, the speed of the cutter is usually adjustable from perhaps as low as 50 rpm to 500 rpm or more.

In operation, the workpiece of rough stock is secured to the milling table, usually by means of clamps. The table may then be adjusted—up or down in a vertical plane, back and forth horizontally, or in a cross-feed manner—to manipulate the work under the cutter. These movements can be made under power by means of rapid-transverse controls, or by duplicate hand controls. In mass-production operations, the table is often manipulated by computer controls that automatically adjust the table to form the desired shape of the workpiece. When precision work is required of only a few pieces, the adjustments are usually made with calibrated hand controls on which each graduation represents .001 in of table movement.

Depending upon the type of milling machine—vertical or horizontal (see Figs. 18.1 and 18.2)—the operator first selects and

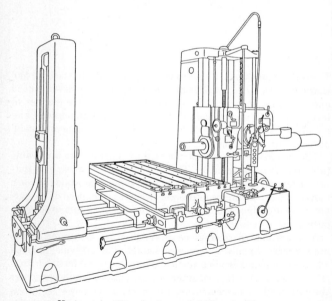

FIG. 18.1 Horizontal milling, boring, and drilling machine.

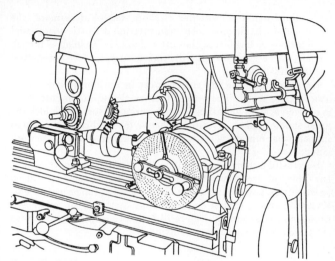

FIG. 18.2   Milling machine set up to mill helical gear teeth.

adjusts the revolving cutter. For a horizontal milling machine, an arbor shaft is fitted into the power-driven spindle. Spacing collars are used to locate the cutter at any position on the arbor, while additional collars and a key hold the cutter in place when the arbor nut is tightened and locked on the arbor to provide a positive drive. A large collar is then set on the end of the arbor to act as a bearing for the overarm bracket, which is slid out and locked in place to support the shaft.

The selection and mounting of the cutter depend on the construction of the machine and the type of work to be performed. The relation of the head stock to the table, however, must first be determined—that is, whether right or left. Furthermore, the direction of cutting movement and rotation of cutter teeth must be determined. In general, the cutter should be as small as the job permits, since a shorter cut will allow for greater feed speeds and requires less power.

On most machines the cutter is set in motion by engaging a quick-acting clutch. The work is then carried up to the cutter and is secured against vibration through various levers which lock the knee, saddle, and cross feed. A selective feed indicator, graduated in inches per minute, is set to govern the speed at which the work is fed.

Successful milling depends upon the flexibility of movement of the table, the accuracy with which the work is adjusted in relation to the cutter, and the use of the correct tools at the proper speeds. Excessive speeds should be avoided, as they dull the cutter and produce a coarse surface. Figure 18.3 shows a gang-mill setup with multiple cutters which speed cutting.

MILLING APPLICATION

Although there are numerous applications to which the milling machine can readily be applied, one example will be explained to demonstrate the practical use of the modern milling machine.

The accurate cutting of keyways on machine gears is one of the many important operations performed by the milling machine. A typical application may call for a keyway $\frac{3}{16}$ in deep by $\frac{5}{16}$ in wide by 3 in long in each end of a 2-in-diameter shaft, 18 in long. To accurately mill such keyways, the operator will first make certain that the milling table is perfectly clean so that all parts that are fitted or clamped together will line up accurately. Then the operating mechanism is lubricated and the shaft to be milled is rigidly

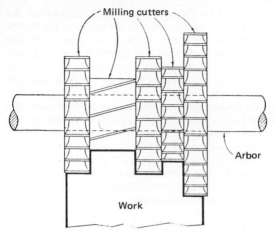

FIG. 18.3 Gang-mill setup for milling machine.

secured in place with two u clamps and heel blocks. An arbor is next inserted in the milling spindle and a draw-in bar is screwed in from the rear to set it solidly in position. Care is taken to see that both the spindle and arbor tapers are clean.

A suitable slotting cutter is selected with teeth of the required width dimension of ⁵⁄₁₆ in. This is placed on the arbor with the necessary collars. The overarm bracket is then positioned to support the arbor, carefully located far enough out to clear the table clamps when the shaft is run in for cutting.

Next comes the centering of the shaft under the cutter. Using the hand controls, the shaft is raised alongside the cutter, leaving a space of about .015 in between the shaft and the teeth of the cutter as determined by a .015-in feeler gage. In this position, the distance from the center of the cutter to the center of the shaft is equal to .15625 in (half the width of the cutter) plus .015 in (the thickness of the feeler gage) plus 1 in (the radius of the shaft). This totals 1.17125 in and represents the distance the table must be moved in toward the head stock in order to center the shaft precisely under the cutter.

To make this adjustment, lower the table and set the index dial on the cross-feed screw at zero. Since the dial is divided into 250 calibrations, each of which represents a table movement of .001 in, a full turn advances the table .250, or ¼ in, and four complete turns plus .171 in will bring the shaft to the desired location.

With all of the preceding operations out of the way, calculating the proper cutting speed comes next. Three factors must be considered when performing this calculation:

1. The nature of the material to be cut.

2. The material from which the cutter is made.

3. The diameter of the cutter.

Let's assume that the material to be cut is cold-rolled steel, and the cutter diameter is 5 in and made of high-speed steel. According to reference charts, 100 surface feet per minute at the cutter teeth is the recommended speed for this particular operation.

Since the cutter diameter is 5 in, the cutter's circumference will be 5 times pi, or 15.708 in; this converts to 1.309 ft. Since the speed in feet per minute at the cutting teeth is to be 100, dividing this speed by the circumference of the cutter in feet gives 76—the number of revolutions needed to obtain the required cutting speed.

The correct feed comes next; that is, the number of inches per minute at which the work is brought into the cutter. For a ³⁄₁₆-in cut (the depth of the keyways) each tooth should remove a chip about .002 in thick. This chip thickness, times the number of teeth on the cutter—say 20—equals .040 in, or the distance the work is fed into the cutter in one complete revolution of the latter. To find the feed per minute, multiply .040 in by the cutter speed of 76 rpm, which gives a feed of 3.04 in/min, or the closest setting to this on the milling machine.

The milling machine is now ready for a short trial cut, at the end of which the milling table is brought up by hand until the cutter just misses the shaft. The machine is then stopped, and the vertical feed index is set at zero. Now, the table is traversed manually until the cutter rests 3 in from the end of the shaft, the specified length of the keyway in this instance. On some machines, a table stop may be utilized to halt the table automatically when it reaches this position; then the table is repositioned to the height required for the cut.

The keyway is to be ³⁄₁₆ in deep, but milling the cylindrical shaft will produce a chord, and it is from this that the depth of the cut must be measured. Consulting a table, the operator finds that the metal removed down to the chord is .0124 in, so the table must be raised .1875 in plus the thickness of the chord segment (.0124 in), which gives .1999 or roughly .200 in. Since one complete revolution of the vertical feed screw is equal to .100 in, two revolutions will provide the desired movement.

The knee of the table is locked in this position. The milling machine is then started and cuts the keyway a length of about ¼ in. The machine is again stopped, and the table is run back so that the depth of the cut can be measured. If the measurement is found to be correct, the milling machine is once again activated and the keyway is finished. If the depth of the cut is off, most adjustments must be made before the cut is completed.

Milling in the Lathe

Milling and keyway cutting attachments for lathes allow the lathe operator to perform many milling applications. These attachments are especially useful in shops that do not have enough work to warrant the installation of an expensive milling machine.

With the lathe milling attachment, the cut is controlled by the handwheel of the lathe carriage, the cross-feed screw of the lathe, and the vertical adjusting screw at the top of the milling attachment. All milling cuts should be taken with the rotation of the cutter against the direction of the feed as shown in Fig. 18.4.

The keyway cutting operation previously described can also be performed on the lathe. The recognized standards for the depth and width of keyways in pulleys, gears, and the like are shown in Fig. 18.5. The same specifications are used for the depth and width of keyways in shafts. In milling a keyway, the key should fit snugly in the keyway but must not be too tight.

Gears may also be cut on the lathe, but a dividing head must be used in conjunction with the milling attachment. Such a combi-

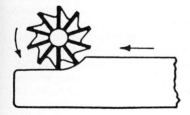

FIG. 18.4 All milling cuts should be taken with the rotation of the cutter against the direction of the feed.

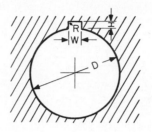

Specifications of American Standard Keyways

Diameter hole D, in	Width W, in	Depth H, in	Radius R, in	Diameter hole D, in	Width W, in	Depth H, in	Radius R, in
½	³⁄₃₂	³⁄₆₄	.020	2½	⅝	⁷⁄₃₂	¹⁄₁₆
⅝ to ⅞	⅛	¹⁄₁₆	¹⁄₃₂	3	¾	¼	³⁄₃₂
1	¼	³⁄₃₂	³⁄₆₄	3½	⅞	⅜	³⁄₃₂
1¼	⁵⁄₁₆	⅛	¹⁄₁₆	4	1	⅜	³⁄₃₂
1½	⅜	⁵⁄₃₂	¹⁄₁₆	4½	1⅛	⁷⁄₁₆	⅛
1¾	⁷⁄₁₆	³⁄₁₆	¹⁄₁₆	5	1¼	½	⅛
2	½	³⁄₁₆	¹⁄₁₆				

FIG. 18.5 Recognized standards for depth and width of keyways.

nation is capable of doing graduating and milling, external key seating, cutting at angles, splining, clotting and all regular dividing-head milling work. With such an attachment, the lathe is practical for cutting small gears and for milling small light work of various kinds. Other uses should become apparent once the basic principles have been learned.

Several types of milling cutters are shown in Fig. 18.6.

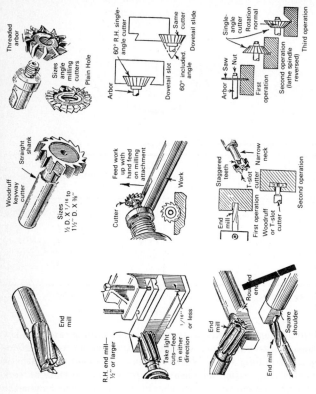

Threaded arbor

Sizes angle milling cutters

Plain Hole

60° R.H. single-angle cutter

Same cutter

Dovetail slide

Arbor

Dovetail slot

60° included angle

Single-angle cutter

Rotation normal

First operation

Arbor

Saw

Nut

Second operation (lathe spindle reversed)

Third operation

Woodruff keyway cutter

Straight shank

Sizes ½ D. X ¹/₁₆ to 1½″ D. X ⅜″

Cutter

Feed work up with hand feed on milling attachment

Work

Staggered teeth

Narrow neck

T-slot cutter

End mill

Woodruff or T-slot cutter

First operation

Second operation

End mill

R.H. end mill— ½″ or larger

Take light cuts—feed in either direction

¹/₁₆″ or less

End mill

Rounded end

End mill

Square shoulder

FIG. 18.6 Several types of milling cutters and their uses.

19

Dividing Heads and Indexing

The making of gears, reamers, and similar items is practical only with a dividing head. When such a dividing attachment is used in conjunction with a milling machine, or a milling attachment on a lathe, many items can be manufactured in the machine shop.

In general, the index head on a milling machine is used to rotate a piece of work through given angles—usually equal divisions of a circle. An indexing head, in combination with the longitudinal feeding movement of the milling table, is used to impart a rotary motion to a workpiece for helical milling action, such as in milling the helical flutes of a cutter.

To operate the index head, the operator rotates the spindle through a desired angle by turning the index crank which controls the interposed gearing. In most dividing or indexing heads, 40 revolutions of the index crank are required for 1 revolution of a spindle. The dividing head can be used for several different methods of indexing, and index tables are provided by the machine manufacturers to aid in obtaining angular spacings and divisions.

The simplest method is called direct or rapid indexing. The spindle is turned through a given angle without interposition of gearing. Plain indexing is an operation in which the spindle is turned through a given angle with the interposition of gearing between the index crank and the spindle. The gearing usually consists of a worm on the index crankshaft which meshes with a worm wheel on the spindle.

COMPOUND INDEXING

This type of indexing is performed by first turning the index crank to a definite setting as in plain indexing, and then turning the index itself to locate the crank in the correct position. Compound indexing can be used to obtain divisions that are required for gun work and goes beyond the range of plain indexing.

DIFFERENTIAL INDEXING

This is a method of indexing where the spindle is turned through a desired division by manipulating the index crank; the index plate is rotated, in turn, by proper gearing that connects it to the spindle. As the crank is rotated, the index plate also rotates a definite amount, depending on the gears that are used. The result is a differential action of the index plate, which can be either in the same direction or in the opposite direction in relation to the direction of crank movement, depending on the gear setup. As motion is a relative matter, the actual motion of the crank at each indexing is either greater or less than its motion relative to the index plate.

In compound indexing, the index plate is rotated manually, with a possibility of error in counting the holes. This is avoided in differential indexing; therefore chances for error are greatly reduced. In the differential indexing operation, the index crank is moved relative to the index plate in the same circular row of holes in a manner that is similar to plain indexing. As the spindle and index plate are connected by interposed gearing, the index plate stop point on the rear of the plate must be disengaged before the plate can be rotated.

ANGULAR INDEXING

This is the operation of rotating the spindle through a definite angle (in degrees) by turning the crank. Sometimes, instead of specifying the number of divisions or sides required for the work to be milled, a given angle, such as 20 degrees or 45 degrees, may be specified for indexing. The number of turns of the index crank required to rotate the spindle 1 degree must first be established to provide a basis for rotating the spindle through a given angle. Usually, 40 turns of the index crank are required to rotate the spindle one complete revolution (360 degrees). Thus, one turn of the crank equals $360/40 = 9$ degrees; or one-ninth turn of the crank rotates the spindle 1 degree. Accordingly, to index 1 degree, the crank must be moved as follows:

1. On a 9-hole index plate, 1 hole.
2. On an 18-hole index plate, 2 holes.
3. On a 27-hole index plate, 3 holes.

BLOCK INDEXING

This type of indexing is sometimes called "multiple indexing" and is adapted mainly for gear cutting. In this operation, the gear teeth are cut in groups separated by spaces; the work is rotated several revolutions by the spindle while the gear teeth are being cut.

The chief advantage of block indexing is that the heat generated by the cutter is distributed more evenly around the rim of the gear; thus, distortion due to local heating is avoided, and higher speeds and feeds can be used.

A Myford 1495 dividing attachment is shown in Fig. 19.1. This attachment is practical for cutting small gears and for milling light work of various kinds on the Myford Super 7 lathe, and will serve to demonstrate how a dividing head is operated. A chart

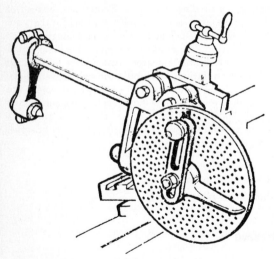

FIG. 19.1 Myford 1495 dividing attachment.

showing the worm wheel ratio of this attachment is shown in Fig. 19.2.

To make the fullest possible use of the dividing attachment, the theory of its operation must be thoroughly understood. As stated in the chart in Fig. 19.2, the ratio of the worm wheel to the worm is 60:1. This means that 60 revolutions of the index crank—which operates the worm—are required for 1 revolution of the spindle. For four divisions, for example, ¼ of 60 is needed; that is, 15 revolutions of the index crank per division:

$$60/4 = 15$$

where 60 is the constant for the attachment, 4 is the number of divisions required, and 15 is the number of revolutions of the index crank (i.e., of the worm) required for each division.

Taking an example from the chart, suppose that 9 divisions are required. We have,

$$60/9 = 20/3 = 6\tfrac{2}{3}$$

Referring to the top left-hand corner of the chart, note that on the no. 1 plate, only the circle with 45 holes is divisible by 3. However, on the no. 2 plate, 42 is also divisible by 3, and could also be used. Considering the former, however, we get

$$60/9 = 20/3 = 6\tfrac{2}{3} = 6\tfrac{30}{45}$$

For each division, six complete revolutions of the index crank plus 30 holes on the 45-hole circle is required.

It will be obvious that, although the chart goes only up to 100, there will be many numbers above this which can be directly obtained from the standard plates. Consider 124 divisions:

$$60/124 = 15/31$$

MYFORD

DIVIDING ATTACHMENT

WORM AND WHEEL RATIO :- 60/1

PLATES 3 AND 4 ARE SUPPLIED AS EXTRA

PLATE Nº1. CIRCLES
91:77:49:45:38:34:32

PLATE Nº2. CIRCLES
47:46:43:42:41:37:31:29

PLATE Nº3 CIRCLES
97:83:73:67:61:27

PLATE Nº4 CIRCLES
89:79:71:66:59:53

Nº OF DIVISIONS	INDEX CIRCLE	Nº OF TURNS OF INDEX CRANK
1	ANY	60
2	"	30
3	"	20
4	"	15
5	"	12

Nº OF DIVISIONS	INDEX CIRCLE	Nº OF TURNS OF INDEX CRANK
21	49	$2\frac{42}{49}$
22	77	$2\frac{56}{77}$
23	46	$2\frac{28}{46}$
24	32	$2\frac{16}{32}$
25	45	$2\frac{18}{45}$

Nº OF DIVISIONS	INDEX CIRCLE	Nº OF TURNS OF INDEX CRANK
41	41	$1\frac{19}{41}$
42	49	$1\frac{21}{49}$
43	43	$1\frac{17}{43}$
44	77	$1\frac{28}{77}$
45	45	$1\frac{15}{45}$

Nº OF DIVISIONS	INDEX CIRCLE	Nº OF TURNS OF INDEX CRANK
61	61	$\frac{60}{61}$
62	31	$\frac{30}{31}$
63	42	$\frac{40}{42}$
64	32	$\frac{30}{32}$
65	91	$\frac{64}{91}$

Nº OF DIVISIONS	INDEX CIRCLE	Nº OF TURNS OF INDEX CRANK
81	27	$\frac{20}{27}$
82	41	$\frac{30}{41}$
83	83	$\frac{60}{83}$
84	49	$\frac{35}{49}$
85	34	$\frac{24}{34}$

No.	Circle	Ratio	No.	Circle	Ratio	No.	Circle	Ratio	No.	Circle	Ratio	No.	Circle	Ratio
6	-	10	26	91	$2\frac{28}{91}$	46	46	$1\frac{14}{46}$	66	77	$\frac{70}{77}$	86	43	$\frac{30}{43}$
7	49	$8\frac{28}{49}$	27	45	$2\frac{10}{45}$	47	47	$1\frac{13}{47}$	67	67	$\frac{60}{67}$	87	29	$\frac{20}{29}$
8	32	$7\frac{16}{32}$	28	42	$2\frac{6}{42}$	48	32	$1\frac{8}{32}$	68	34	$\frac{30}{34}$	88	66	$\frac{45}{66}$
9	45	$6\frac{30}{45}$	29	29	$2\frac{2}{29}$	49	49	$1\frac{11}{49}$	69	46	$\frac{40}{46}$	89	89	$\frac{60}{89}$
10	ANY	6	30	ANY	2	50	45	$1\frac{9}{45}$	70	49	$\frac{42}{49}$	90	45	$\frac{30}{45}$
11	77	$5\frac{35}{77}$	31	31	$1\frac{29}{31}$	51	34	$1\frac{6}{34}$	71	71	$\frac{60}{71}$	91	91	$\frac{60}{91}$
12	ANY	5	32	32	$1\frac{28}{32}$	52	91	$1\frac{14}{91}$	72	42	$\frac{35}{42}$	92	46	$\frac{30}{46}$
13	91	$4\frac{56}{91}$	33	77	$1\frac{63}{77}$	53	53	$1\frac{7}{53}$	73	73	$\frac{60}{73}$	93	31	$\frac{20}{31}$
14	49	$4\frac{14}{49}$	34	34	$1\frac{26}{34}$	54	45	$1\frac{5}{45}$	74	37	$\frac{30}{37}$	94	47	$\frac{30}{47}$
15	ANY	4	35	49	$1\frac{35}{49}$	55	77	$1\frac{7}{77}$	75	45	$\frac{36}{45}$	95	38	$\frac{24}{38}$
16	32	$3\frac{24}{32}$	36	45	$1\frac{30}{45}$	56	42	$1\frac{3}{42}$	76	38	$\frac{30}{38}$	96	32	$\frac{20}{32}$
17	34	$3\frac{18}{34}$	37	37	$1\frac{23}{37}$	57	38	$1\frac{2}{38}$	77	77	$\frac{60}{77}$	97	97	$\frac{60}{97}$
18	45	$3\frac{15}{45}$	38	38	$1\frac{22}{38}$	58	29	$1\frac{1}{29}$	78	91	$\frac{70}{91}$	98	49	$\frac{30}{49}$
19	38	$3\frac{6}{38}$	39	91	$1\frac{49}{91}$	59	59	$1\frac{1}{59}$	79	79	$\frac{60}{79}$	99	66	$\frac{40}{66}$
20	ANY	3	40	32	$1\frac{16}{32}$	60	ANY	1	80	32	$\frac{24}{32}$	100	45	$\frac{27}{45}$

FOR ANGULAR DIVIDING USE 42 CIRCLE $1° = 42$

FIG. 19.2 Worm wheel ratio for attachment in Fig. 19.1.

The no. 2 plate has a circle having 31 holes so that, using this, it is necessary to advance the index crank 15 holes for each division.

It will be realized that a further range of divisions can be obtained if the user is prepared to make extra plates. Take for example 144 divisions:

$$60/144 = 5/12$$

None of the plates has a circle with 12 holes, nor any number divisible by 12, but a plate having this number could readily be made. It may, however, be considered worthwhile to have 24 holes, since this would enable 288 divisions to be obtained if required at a later date.

THE SECTOR ARMS

The two arms of the assembly are secured, one to the other, in any required setting by a slotted screw. The whole assembly, which is mounted on the worm shaft bracket, is retained in position by friction, so that, while it can be readily rotated as the dividing operation progresses, it is firmly held in the individual settings.

Where the movement of the index crank is less than about two-thirds of a revolution per division (see example for 124 divisions explained previously) or a whole number of revolutions plus a portion of a revolution less than about two-thirds of a revolution (see chart in Fig. 19.2 for 25 divisions), the movement of the plunger on the index crank will be between the arms.

Where the arc of movement of the index crank exceeds roughly two-thirds of a revolution or a number of full turns plus more than about two-thirds of a revolution, the plunger will operate over the sector arms (see chart for 16 divisions).

Setting the sector arms for operation between the arms (e.g., 25 divisions), the following instructions should be followed.

Mount the no. 1 division plate with the numbered holes to the top. Adjust the setting of the index crank so that the plunger will drop into the holes in the 45-hole circle. Open out the sector arms so that there is about 180 degrees between them. With one arm (A) to the *left* of the numbered hole and the other arm (B) at about 4 o'clock, rotate the crank *clockwise* until the plunger will drop into the numbered hole. Next, rotate the sector arms clockwise until arm A is touching the left-hand side of the plunger and arm B is at about 6 o'clock. Unless it is already free, slacken the screw which clamps the two arms together.

Starting at the hole to the *right* of the plunger, count 18 holes in a clockwise direction and, holding arm A firmly against the left-hand side of the plunger, move arm B counterclockwise until its inner edge lines up with the lower edge of the eighteenth hole and tighten the screw to lock the sector arms together. (In this example, there should be a total of 19 holes between the sector arms.)

The following instructions are for setting the sector arms for operation over the arms (e.g., 16 divisions).

Open up the sector arms so that there is about 120 degrees between them. Adjust the setting of the index crank so that the plunger will drop into the holes of the 32-hole circle. (no. 1 plate). With arm B to the *right* of the numbered holes and arm A at about 8 o'clock, rotate the crank *clockwise* until the plunger will drop into the numbered hole. Next rotate the sector arms *counterclockwise* so that arm B which is to the right of the plunger is touching it (still on the right-hand side).

Starting at the hole to the left of the plunger count 8 (i.e., 32 − 24) holes in a counterclockwise direction and, holding arm B firmly against the right-hand side of the plunger, move arm A upward in a clockwise direction until its upper edge lines up with the lower edge of the required hole. Tighten the screw to lock the sector arms together (in this example there will be a total of 9 holes between the sector arms).

MOUNTING THE ATTACHMENT

Where circumstances permit and both slides are available, it will generally be found preferable to use the plain vertical slide 67/1, as this will give greater rigidity. In addition, the maximum height of the attachment spindle above the head stock spindle is slightly greater than on the swiveling slide 68/2.

When working with the axis of the attachment spindle parallel to the axis of the head stock spindle, it will generally be preferable to have the attachment spindle at the same height as the head stock spindle and thus work on the horizontal centerline.

When using a cutter on an arbor between centers, with the axes of dividing attachment and head stock spindles at right angles one to the other, if the sum of the radii of cutter and work exceeds 2⅛ in (e.g., a 2-in-diameter cutter and a 2½-in-diameter component), the dividing attachment may be raised 2⅛ in by using a no. 30/011 raising block for the Myford Super 7 lathe. Where this is not available, or if the sum of the radii exceeds 4¼ in, it will be necessary to increase the center distance between cutter and work in some other way.

This may be done by repositioning the key in the back of the spindle bracket by fitting it into the vertical slot, rotating the table of the swiveling slide 68/2, and mounting the attachment with the slide so swiveled that the attachment spindle is looking up at an angle. The axis of the feed screw of the milling slide will then be parallel to the axis of the spindle of the dividing attachment. The approximate setting for height may be made with both cutter and workpiece in position by rotating the slide table and dividing attachment and simultaneously advancing or withdrawing the cross slide, and, by means of its feed screw, adjusting the position of the milling slide table.

The workpiece will be fed across the cutter by the feed screw of the swiveling slide. The depth of cut will be controlled by the cross-slide feed screw. It will be noted, however, that the depth of cut actually applied will be less than the movement of the cross slide and that the relationship will vary with differing angles.

The amount of cross-slide movement for any required depth of cut may be calculated by using the formula

$$f = \frac{c}{\sin A}$$

where f = amount cross slide is to be moved
c = required depth of cut
A = angle of inclination of slide and dividing attachment

Reference to trigonometric tables will show that an inclination of 30 degrees to the horizontal should be chosen whenever possible since the amount of movement of the cross slide will be exactly twice the required depth of cut.

Where the workpiece is suitable it should be mounted in a chuck or no. 1031 collet with 1438 nosepiece. It is not advised that the work be mounted between centers, as the outboard support is not intended for this purpose. Components which are bored should be mounted on a stub arbor machined to suit. This in turn may be mounted in a chuck, in the no. 1031 collet, or it may have a no. 2 Morse taper shank and be mounted directly in the taper in the spindle of the attachment.

If the component has a center at its outer end or is mounted on a stub arbor with such a center, the outboard center may be used for additional support.

A representative indexing operation is cutting of the teeth of a 25-tooth spur gear. First, mount the cutter on an arbor between centers (or held in the chuck, with tail stock support). Then, mount the vertical slide on the lathe cross slide, mount the dividing attachment on the vertical slide, and mount the workpiece. Centralize the workpiece to the cutter and set for height so that the cutter just touches the periphery of the workpiece. Then fit the number one division plate and set the index crank and the sector arms as described previously. Set for the required depth of cut and rotate the index crank clockwise until the plunger drops into the numbered hole. (It may be found convenient when rotating the index crank to stop just short of the required hole and then to tap the index crank around until the plunger drops in).

Make the first cut. Check that the sector arms are in their correct position (arm A touching the left-hand side of the plunger) and withdraw the plunger. Rotate the knob to allow the shallow notch to engage the stop peg. Now rotate the index crank clockwise two complete revolutions plus 18 holes—that is up to arm B—and allow the plunger to drop into the hole in the index plate.

Cut the second tooth. Rotate the sector arms clockwise so that arm A is again touching the plunger and withdraw the plunger and index as before.

When operating over the sector arms (e.g., 16 divisions) the movements of the index crank will again be in a clockwise direction, although this time the sector arms will be moved counterclockwise. However, at the beginning, with the plunger in the numbered hole, arm B will be touching the right-hand side of the plunger, whereas arm A will be at 9 o'clock. The movement of the index crank will be as follows.

Withdraw the plunger, rotate the index crank clockwise three complete revolutions plus 24 holes, that is, just over arm A. Before the next movement of the index crank, rotate the sector arms counterclockwise so that arm B is in contact with the plunger.

If two or more cuts are to be made in achieving full depth, the first, roughing, cut should be made right around the blank before adjusting for depth for the subsequent cuts.

As you use the dividing attachment, you will gain the experience necessary for perfect operation, but here are a few tips in the meantime. When operating the attachment, always rotate the index crank clockwise and if you inadvertently go beyond the required hole do not come directly back to it. Go back well past it so that you can again come up to it in a clockwise direction.

The bore for the main spindle of the attachment is split, the resulting slot being fitted with a packing piece. This is so relieved that the center screw can be used, if required, as a locking screw for extra rigidity during machining. The two outer screws are for bearing clearance adjustment and, having been carefully set during manufacture, should not normally be touched.

ADJUSTMENTS

No adjustments should be necessary until the attachment has had an appreciable amount of use. There are, however, four which can be made:

1. Should the spindle bearing clearance eventually need adjustment, it will be necessary to remove and thin the packing piece.

2. End float in the spindle can be eliminated by adjusting the collar at the rear end of the spindle. This is threaded (right-hand) onto the spindle and is secured by means of a grub screw which must be released for the adjustment to be made and retightened afterwards.

3. Backlash between worm and worm wheel can be eliminated by releasing the hexagon-head screws which secure the worm

bracket to the main body of the attachment and lowering the bracket. Note that the bracket is keyed to the body.

4. The position of the plunger in the index crank can be adjusted if, with the plunger located so that the shallow notch is engaged with the stop peg, the plunger will not pass over the sector arms (allow about $\frac{1}{16}$ in, 1.5 mm, clearance). To adjust, release the socket setscrew in the end of the plunger knob, release the plunger in order to reduce the load on the spring, but hold the knob so that the plunger is not fully engaged with the dividing plate and screw the plunger into the knob the requisite amount. Retighten the socket setscrew to lock the plunger in position in the knob.

20

Arbors, Collets, and Adapters

Cylindrical work that has been bored and reamed in a chuck is usually further machined on a mandrel between lathe centers. The mandrel is slightly tapered and must be driven into the hole tight enough so that the work will not slip on the mandrel while it is being machined.

Large-diameter work such as pulleys should be driven with a pin or driver attached to the lathe faceplate if it can be arranged, as this will eliminate the possibility of the work slipping on the mandrel while it is being turned. Before driving the mandrel into the hole in the work, however, oil both the mandrel and the hole so that the work will be easy to remove when completed. If there is no lubricant on the mandrel it may "freeze" in the work, in which case it cannot be driven out without ruining both the work and mandrel.

When driving a mandrel out of a piece of work, be sure that it is driven in the opposite direction from that in which it entered to take advantage of the taper.

Standard lathe mandrels can be purchased in various sizes, and each is hardened and tempered with a taper of about .006 in/ft. However, in the case of special applications having an odd diameter hole, a soft mandrel may be used, turning and filing it to the proper diameter and tapered for a driving fit in the hole in the work.

Special types of mandrels are often used for special applications. For example, a nut mandrel for finishing the outside diameter of gear blanks is shown in Fig. 20.1. Expansion mandrels of various types are also available and are used where there is considerable variation in the hole sizes.

To make lathe mandrels, cut tool-steel stock to the required length and center-drill. This of course must be done with the utmost accuracy since it is necessary that the outer shell left from the rolling process be uniformly turned off. Mount the piece between centers and rough-turn. At this point, the work should be pack-annealed to eliminate any internal stresses. Afterwards

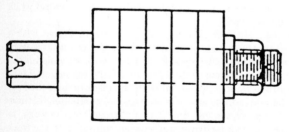

Nut mandrel for finishing
gear blanks

FIG. 20.1

return it to the lathe and bring it to size with several light cuts, leaving it about 0.0002 in oversize to allow for grinding.

Ream and counterbore the holes in the ends very carefully, as the accuracy of the mandrel depends upon their being concentric with the outer diameter. The flat on each end for the set-screw of the lathe dog can be carefully filed or cut with an end mill.

As mentioned previously, mandrels are usually tapered—about 0.006 in/ft—and always have the size stamped on the large end. After it has been marked accordingly, the mandrel can be hardened and tempered, then mounted between centers and ground to the proper taper and final size with the tool-post grinder. Dimensions for solid lathe mandrels are shown in Fig. 20.2.

Expanding mandrels may be made in a similar fashion. In general, the expanding mandrel consists of a tapered arbor and a split bushing taper-bored to fit it. This bushing is turned parallel on the outside, and expands evenly when the arbor is driven into it. Several bushings of different outside diameters may be used with the same arbor. For heavy duty, both parts should be of tool steel, hardened and spring-tempered. For occasional use, unhardened machine steel will serve.

To make the arbor, center-drill the stock for the arbor carefully and then turn it to the required dimensions. To make the bushing, chuck a short steel bar and drill a hole of the required size through it. Bore out the taper to match that of the arbor.

Drive the arbor into the bushing to turn the outside to size, then lay out six evenly spaced holes at each end of the bushing to be center-punched and drilled.

Slots for this type of arbor may be cut with a milling attachment and cutter mounted in the lathe. A steel plug in the bushing holds it in shape during this operation.

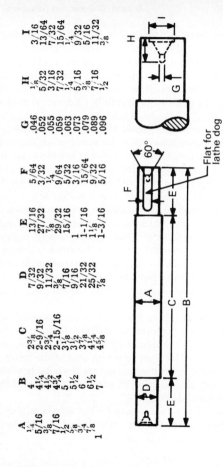

A	B	C	D	E	F	G	H	I
1/4	4	2 3/8	7/32	13/16	5/64	.046	1/8	3/16
5/16	4 1/4	2-9/16	9/32	27/32	3/32	.052	5/32	13/64
3/8	4 1/2	2 3/4	11/32	7/8	1/4	.055	3/16	7/32
7/16	4 3/4	2-15/16	3/8	29/32	9/64	.059	7/32	15/64
1/2	5	3 1/8	7/16	15/16	5/32	.063	1/4	1/4
5/8	5 1/2	3 3/8	9/16	1	3/16	.073	5/16	9/32
3/4	6	3 7/8	21/32	1-1/16	15/64	.079	3/8	5/16
7/8	6 1/2	4 1/4	25/32	1 1/8	9/32	.089	7/16	11/32
1	7	4 5/8	7/8	1-3/16	5/16	.096	1/2	3/8

FIG. 20.2 Dimensions for solid lathe mandrels.

COLLETS

The draw-in collet chuck is one of the most accurate of all types of chucks and is used for precision work in many machine shop tools. These collets are made for round, square, and other shapes. In general, the work to be held in these collets should not be more than .001 in smaller or larger than the collet size. If the diameter of the work varies more than this, it will impair the accuracy and efficiency of the collet. A separate collect should be used for each diameter of the work.

The construction of the draw-in collet chuck is shown in Fig. 20.3. The hollow drawbar with handwheel attached extends through the head stock spindle of the lathe and is threaded on the right end to receive the spring collet. Turning the handwheel to the right draws the collet into the tapered closing sleeve and tightens the collet on the work. Turning the handwheel to the left releases the collet.

The spring collet may be replaced with a stop chuck and closer for holding discs such as gear blanks. A pot collect may be used in place of the stop chuck for small-diameter work.

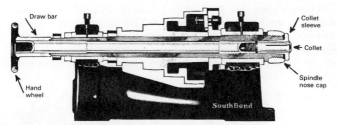

FIG. 20.3 Cross-section of headstock showing construction of draw-in collet chuck attachment.

Another type of collet is the hand-lever draw-in collet chuck. It is similar to the handwheel-type draw-in collect chuck except that the collet is opened and closed by moving the hand lever to the right or left. This permits gripping or releasing the work without stopping the lathe spindle if desired.

There are two important rules for the use of draw-in collet chuck attachments: absolute cleanliness and the selection of the proper size collet. The collets, tapered sleeve, and the inside of the spindle nose must be wiped clean and dry. A collet attachment is the most accurate type of precision chucking and must be treated with the greatest care.

Some different size collets are shown in Fig. 20.4.

LATHE ADAPTERS

Many types of lathe work cannot be machined on centers or in a chuck in a conventional manner; these pieces must be secured for machining in other ways. Many times the work can be secured to the faceplate with bolts, studs, or clamps. In fact, some of the most accurate tool and die operations are handled in this way.

An angle plate (Fig. 20.5) can be bolted to any point on the faceplate for machining irregular shapes and for off-center drilling and boring. However, when heavy pieces are mounted off center, always bolt a counterbalance of equal weight on the opposite edge of the faceplate. The counterbalance protects lathe accuracy by

FIG. 20.4   A set of collets for round work ¹⁄₁₆ to ¾ in by sixteenths.

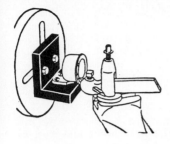

FIG. 20.5 Using the angle plate.

equalizing pressure on the bearings and reduces excessive vibration caused by out-of-balance turning.

The four-jaw independent chuck is another method of holding odd-shaped work. The four jaws on this type of chuck may be adjusted separately and are reversible so that work of any shape can be clamped from the inside or the outside. Some independent chucks are threaded to fit directly on the spindle nose, others are bolted to an adapter plate which fits the spindle.

Mounting work in the four-jaw chuck is largely a matter of centering. Determine the portion of the rough work that is to run true, then clamp the work as closely centered as possible, using as a guide the concentric rings on the face of the chuck. Test for trueness, marking the high spots with chalk rested against the tool post. The chuck jaws should be adjusted until the chalk or tool bit contacts the entire circumference of the work.

If especially accurate centering is desired, the trueness of the work should be checked with the tail stock center by means of an instrument called a center tester.

21

Grinders and Grinding

ABRASIVES

Grinding Wheels

Grinding is the cutting action of thousands of sharp abrasive grains on the face of the grinding wheel. The grain actually cuts chips out of the work. In general, there are two types of grinding methods: offhand grinding and precision grinding. In the first method (offhand grinding), the grinding is done to broad tolerances by either applying the grinding wheel manually to the work or applying the work offhand to the grinding wheel. Applications include snagging castings, weld grinding, some tool sharpening, and miscellaneous rough grinding.

Precision grinding includes machine grinding where the size limits (tolerances) are exceedingly small. Applications will include cylindrical grinding, internal grinding, surface grinding, and tool and cutter grinding.

During the grinding operation, two components are used: the abrasive which does the actual cutting, and the bond, which supports the abrasive grains while they cut.

The perfect grinding wheel is one on which the abrasive grains will break loose from the bond just as they become too dull to cut, and expose new sharp grains underneath. This process repeats itself until the wheel is gone or becomes so badly worn that it is of no practical use.

Wheels for grinding usually are of two types: hard and soft. The soft wheel is used for hard materials since these will dull the abrasives faster, so the grain must break away faster to expose new cutting abrasives.

The hard wheel is used for soft materials, as the grains stay sharper longer and should stay in place long enough to yield reasonable wheel life from it before they break loose.

Aluminum oxide is used for grinding carbon steel, alloy steel, malleable iron, wrought iron, hard bronze, and similar metals, and comes in a broad range of densities to cover a broad range of grinding applications.

Silicon carbide is used for grinding gray iron, chilled iron, brass, soft bronze, copper, aluminum, stone, marble, rubber, hard face alloys, and cemented carbides. There are two classifications:

1. Gray, used for general grinding, heavy-duty snagging, and cylindrical, centerless, and internal grinding of the above materials.

2. Green, generally preferred by operators for grinding cemented carbide tools including masonry bits, percussion carbide hammer bits and other tools using carbide tips.

Diamond grinding wheels are also used in grinding cemented carbides (router bits, etc.), quartz, glass, gem stones, marble, granite, and ceramics.

The grit size of all abrasives represents the approximate number of openings per lineal inch in the final screen used to size the grain. The lower the number, the larger the grit (stone).

The grade indicates the relative strength (holding power) of the bond which holds the abrasive grains in place. There are five types of bond in current use: vitrified, resinoid, rubber, shellac, and silicate.

Coated Abrasives

Coated abrasives are individual cutting tools bonded onto a fiber, cloth, or paper backing. They are commonly referred to as "sandpaper."

Coated abrasives have become a vital production tool, and depended upon by modern industry for increasing production, improving quality of finishes, and reducing costs.

Coated abrasives are products manufactured from three basic raw materials: a mineral, a backing, and an adhesive bond. Minerals do the job of grinding, finishing, and polishing. Five are generally in use:

Flint.   This mineral is actually quartz and is white, resembling white sand.

Emery.   This is dull black and is hard, rounded, blocky.

Garnet.   This is reddish-brown and is of medium hardness with good cutting edges. It has a tendency to break or refracture when in use, forming new cutting edges.

Aluminum oxide.   A brown mineral that is extremely tough, durable, and resistant to wear and is capable of penetrating almost any surface. It stands up without appreciable refracture or breakdown in use.

Silicon carbide. This mineral is shiny black. It is brittle and fractures into sharp, silvery wedges having cutting edges that permit rapid stock removal.

Backings are the base for the mineral and are classified into four basic groups:

Cloth. Four weights of cloth are most commonly used: J, X, Y, and M. Cloth backings are used in both hand and mechanical sanding operations. Open mesh screen-like fabrics are used on Fabricut products.

Paper. There are five weights of paper backings, designated A, C, D, E, and F, A being the lightest weight and F being the heaviest.

Fiber. Fiber is made from a rag-stock paper which has been condensed and hardened and when finished is strong, tough, relatively hard and has much more body than any of the other backings. It is used for discs and for drum sanding materials.

Combination. Two kinds are commonly used: paper and cloth laminated together and fiber and cloth laminated together. These types of backing are sturdy and shock-resistant. The paper-and-cloth types is used mainly on high-speed drum sanders while the fiber-and-cloth type is used for discs.

Coated abrasives have two adhesive layers which anchor and lock the mineral to the backing. The first is called the "make coat" and the second the "size coat." There are five types of adhesive bonds used: glue, glue and filler, resin over glue, resin over resin, and waterproof.

Mineral Grading

There are 22 different grit sizes or grades, as they are commonly called, ranging from the size of this letter O,—which is the coarsest, to a powder-like form, which is the finest. In the coated abrasive industry, two marking systems are used to denote a given size grit: the grit symbol and mesh number.

The grit symbol is the oldest system of marking and runs from 4½ (the coarsest) to ¹⁄₀, which is the finest. The grade number is now used instead of the grit symbol. The grade number, or mesh number, refers to the number of openings to a lineal inch on a standard control screen. The mesh number markings range from no. 12 (the coarsest) to no. 600 (the finest).

Simplified markings are used on some abrasive papers. They are "Extra Fine," "Fine," "Medium," "Coarse," and "Extra Coarse."

Coatings

There are two basic types of abrasive coatings, known as closed coat and open coat. In the closed coat, the abrasive grains are adjacent to one another without voids between so that the backing is completely covered with abrasive grain. In the open coat, the abrasive grains are set at a predetermined distance from one another with a void between. The surface coverage by the abrasive grain is about 50 to 70 percent.

Because of the ever-expanding application of coated abrasives as production tools, they are used at some point in the manufacture of almost all fabricated articles.

Manufacturers, processors, and fabricators use coated abrasives from the widest range of grinding, finishing, and polishing—from high-volume stock removal or grinding the toughest welds to developing a high-lustre finish on stainless steel. They are used in every metalworking operations from large industrial plants to small machine shops.

Lubricants

Following is a list of common lubricant types:
Water
Water solutions
Water-soluble oil
Straight mineral oil
Straight fatty oils
Mineral lard oil
Sulfurized and chlorinated cutting oils
Wax
Grease sticks

Storing Abrasives

Proper storage of abrasives will preserve their factory quality and cutting power. Since coated abrasives are affected by widespread changes in atmospheric conditions, the following precautions and recommendations are suggested:

1. Avoid too much moisture or extremely dry locations. The ideal relative humidity is between 35 and 50 percent.

2. Avoid very hot or cold temperatures. A constant temperature of 70°F is recommended.

3. Do not store near radiator, warm-air register, or open window.

4. Do not expose sandpaper to sunlight.

5. Store sandpaper on racks. Keep in original shipping container until ready to use.

Mounted Stones

Mounted wheels are designed for use in high-speed die grinders and straight grinders. While the wheels are formed on mandrels of several diameters, the most common are ⅛- and ¼-in diameter. These mandrels have knurled ends which are securely anchored into the abrasive wheel with special high-strength cements to help eliminate separation. The mounted wheel is then ground to run concentric with its mandrel, resulting in a precisely finished product which is ready to be used.

Combinations of abrasive grains with vitrified and resinoid bonds provide the various grinding actions normally required of mounted wheels. Several basic types of aluminum oxide and silicon carbide abrasive are combined in a large range of grain sizes with vitrified and resinoid bonds in various densities to obtain the desired results.

It is important to check maximum wheel diameters recommended for the grinder before using. Improper wheel use can cause injury. To avoid injury, do not use wheels that have been dropped, store wheels properly when not in use, and true up wheels when needed. Table 21.1 lists various types of wheels and their maximum operating speeds.

Special Abrasives

The cleaning and finishing of metal surfaces is a subject of increasing interest in a time of escalating production cost, pollution restriction, and consumer demand for high quality and longer product life. New requirements for better surface preparation have led to improved cleaning and finishing materials. 3-M Company has introduced a product called Scotch-Brite three-dimensional abrasives to meet these needs. They do not undercut, or change the shape of a workpiece. They are for cleaning and con-

TABLE 21.1   Grinding Wheel Speeds

Wheel diameter		Peripheral speed, ft/min (m/min)					
in	mm (approx.)	4000 (1219)	4500 (1372)	5000 (1524)	5500 (1676)	6000 (1829)	6500 (1981)
		Revolutions per minute					
¼	6	61,116	68,756	76,392	84,032	91,672	99,212
⅜	10	40,744	46,594	50,928	56,021	61,115	66,141
½	13	30,558	34,378	38,196	42,016	45,836	49,656
⅝	16	24,446	27,502	30,557	33,615	36,669	39,685
¾	19	20,372	22,918	25,464	28,011	30,557	33,071
⅞	22	17,462	19,645	21,826	24,009	26,192	28,346
1	25	15,279	17,189	19,098	21,008	22,918	24,828
2	51	7,639	8,594	9,549	10,504	11,459	12,414
3	76	5,093	5,729	6,366	7,003	7,631	8,276
4	102	3,820	4,297	4,775	5,252	5,729	6,207
5	127	3,056	3,438	3,820	4,202	4,584	4,966
6	152	2,546	2,865	3,183	3,501	3,820	4,138
7	178	2,183	2,455	2,728	3,001	3,274	3,547
8	203	1,190	2,148	2,387	2,626	2,865	3,103
10	254	1,528	1,719	1,190	2,101	2,292	2,483
12	305	1,273	1,432	1,591	1,751	1,190	2,069
14	356	1,091	1,228	1,364	1,500	1,637	1,773
16	406	955	1,074	1,194	1,313	1,432	1,552
18	457	849	955	1,061	1,167	1,273	1,379
20	508	764	859	955	1,050	1,146	1,241
22	559	694	781	868	955	1,042	1,128
24	610	637	716	796	875	955	1,034
26	660	588	661	734	808	881	855
28	711	546	614	682	750	818	887
30	762	509	573	637	700	764	828
32	813	477	537	597	656	716	776
34	864	449	505	562	618	674	730
36	914	424	477	530	583	637	690

SOURCE: The L. S. Starrett Co.

ditioning metal surfaces in preparation for applying coatings, metal plating, or laminating film. They remove normal oxide film on all metal surfaces and remove dirt and residual contaminants.

Special "Clean 'n Strip" discs and wheels can be used like wire brushes to clean metal surfaces, welds, injection molders, etc. They eliminate the problem of loose flying wires.

Scotch-Brite abrasives move along the surface of a workpiece in a spring-like fashing. Burrs and contaminants are removed without appreciably altering the shape of the workpiece. This is especially important when working with irregular or contoured pieces.

Scotch-Brite abrasives are made up of a nonwoven synthetic-fiber web, impregnated throughout with abrasive particles that are held in place by high-strength resins.

Because the abrasive particles are uniformly impregnated throughout the open web, fresh grain is continually exposed as old grain wears away. No loading occurs. The synthetic fibers will not rust or cause metal contamination. they can be used dry or with water, oils, or mild cleansers.

PRECISION GRINDING

Precision grinding is used in the machine shop for the final finish on certain classes of work where the closest of precision and perfection of finish is required. A tool-post grinder (Fig. 21.1) is the tool normally used for such grinding operations.

During precision grinding, the following general requirements must be met to obtain an adequate finish on the workpieces;

1. The grinding wheel must be kept clean at all times; it should be dressed either with a diamond or a piece of carborundum.
2. Never take big cuts during this final stage of finishing—usually

FIG. 21.1 Typical tool post grinder used on metal turning lathe. *(Atlas)*

no larger than .002 in per pass. On the first pass, just run the grinding wheel over the surface to be ground, without taking any cut at all.

3. The softer the material being ground, the more the care that must be taken so as not to remove too much metal per pass.

4. Rigidity of support both for the work and the grinder are essential. An unsteady support or grinder will leave unsightly waves on the work and will eliminate all possibility of accuracy.

In most cases, precision grinding is applied primarily to metals, but there are many instances where this operation can be used to an amazing degree of accuracy and perfection on plastics also.

GRINDING IN THE LATHE

A lathe equipped with a grinding attachment, such as a tool post grinder, can perform all types of grinding with a high degree of accuracy. The lathe can then be used for sharpening reamers and milling cutters. grinding hardened bushings and shafts, and many other grinding operations.

During any grinding operation in the lathe, the ways should be protected with paper, oilcloth, or canvas, as grit from the grinding wheel is extremely harmful to the lathe ways. The lathe spindle bearings should also be protected. A small pan of oil or water placed just below the grinding wheel will collect most of the dust and grit.

A grinder for external grinding should have a wheel at least 4 in in diameter. The grinder should be mounted directly on the compound rest of the lathe.

For internal grinding, a high-speed internal grinding attachment is normally used. A spindle speed of 30,000 rpm or more is possible with these high-speed grinders.

Dressing Grinding Wheel

The grinding wheel must be balanced and must be dressed with a diamond dresser if a smooth, accurate ground finish is to be obtained. The grinding wheel must be dressed frequently; in fact, it should be dressed before every operation and in exactly the position in which it will be used.

The diamond dresser consists of a small industrial diamond mounted in a steel shank. The dresser must be rigidly supported in a fixture for turning the grinding wheel, as shown in Fig. 21.2. The diamond point of the dresser should be placed on center, or slightly below center, and the revolving grinding wheel passed lightly back and forth across the diamond. Remove about .001 in from the wheel at each pass and dress the wheel just enough to make it run true and remove all glazed surfaces from the wheel.

Grinding Hardened Steel Parts

Hardened steel parts should be carefully ground in order to produce a smooth, accurate finish. The part should be machined to within a few thousandths of the finished size before it is hardened. After hardening, all scale should be removed before grinding. Remove only a few thousandths at each pass of the grinding wheel because if the part is ground too fast it may become overheated and warp, or the temper may be drawn.

In grinding operations, the work must turn in a direction opposite that of the grinding wheel. Figure 21.3 shows how the rotation

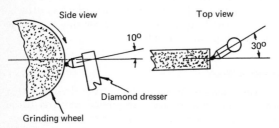

FIG. 21.2 Proper position of diamond point for dressing grinding wheel.

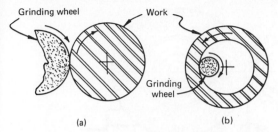

FIG. 21.3 For precision grinding in the lathe, the work must always turn in a direction opposite to that of the grinding wheel. (*a*) External grinding and (*b*) internal grinding.

of the lathe spindle must be clockwise (backward) for external grinding and counterclockwise (forward) for internal grinding.

Sharpening Reamers and Cutters

Reamers and milling cutters may be sharpened by grinding in the lathe. Some reamers are first circular-ground, then relieved by grinding with a tooth rest set slightly below center as shown in Fig. 21.4, leaving a land .002 to .005 in wide. Other reamers and most milling cutters are ground with about 2 degrees relief.

EXTERNAL GRINDING

Since grinding is a finishing operation, the work should be turned as close to the final finish size as possible before the grinding operation is begun.

With the work and the grinder mounted in position and the

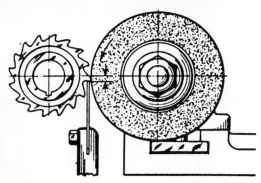

FIG. 21.4 Grinding clearance on a milling cutter. (*South Bend Lathe Co.*)

grinding wheel dressed properly, advance the wheel into one end of the work. Take light cuts across the entire length of the work. If using the automatic carriage feed, set up the change gears for the .0035-in feed. Hand feeds should be very slow and even. The last finishing cut should be less than .001 in; in many cases, a last cut is taken without advancing the feed. When hardened stock is being ground, redress the wheel before taking the final cuts.

INTERNAL GRINDING

The exact method of internal grinding will depend upon the type of grinding attachment. However, in a typical attachment, a quill is used to hold the grinding wheel for internal grinding and is threaded and tapered to fit inside the grinder spindle after the external wheel is removed. The higher grinding speeds should be used, and the lathe spindle must be rotating in a forward direction.

During internal grinding operations, it is necessary to take light cuts and feed in very slowly because of the overhand of the grinding wheel. After the last cut allow the wheel to pass back and forth across the work several times without advancing the feed.

GRINDING VALVES

Mount the valve in a Jacobs head stock chuck and use the external wheel on the grinder spindle. Align the spindle with the tail stock center, and dress the wheel with the compound rest set at the proper angle for the valve in question (usually 15 to 75 degrees). With the lathe spindle turning in a direction opposite that of the grinding wheel, feed in slowly with the compound rest feed, taking light cuts.

GRINDING FLAT VALVES

Mount the valve in the Jacobs head stock chuck. Mount the internal wheel on the grinder spindle and align the spindle with the tail stock center. With the compound rest set at 0, dress the grinding wheel, feeding across the diamond dresser with the compound rest feed. With the lathe spindle turning in a direction opposite that of the grinding wheel, feed in slowly with the compound rest feed, taking light cuts.

GRINDING 60 DEGREE LATHE CENTERS

With the center and sleeve mounted in the lathe spindle, dress the external grinding wheel with the compound rest set at exactly the proper angle. Adjust the speed to obtain a slow lathe spindle speed and shift the reversing switch lever so that the lathe spindle is turning in a direction opposite that of the grinding wheel. Feed up to the center with the carriage handwheel. Lock the carriage in position and feed across the center slowly with the compound rest feed, taking light cuts.

SURFACE GRINDING

Precision surface grinding operations are an important part of modern machine-shop practice. The accuracy of jigs, fixtures, dies, and other special tools and machines is dependent upon the accurate surface grinding of the various parts that enter into their construction. The following machines are those that are currently in common use:

1. Universal tool and cutter grinders arranged for face grinding, or for light surface grinding with a straight wheel and reciprocating table.
2. Reciprocating table, using a straight wheel mounted on a horizontal spindle.
3. Reciporcating table, using a cup or cylinder wheel mounted on a vertical spindle.
4. Rotary table, using a striaght wheel mounted on a horizontal spindle.

TABLE 21.2 Allowances for Grinding

Diameter, in	Length, in Allowance, in										
	3	6	9	12	15	18	24	30	36	42	48
½	.010	.010	.010	.010	.015	.015	.015	.020	.020	.020	.020
¾	.010	.010	.010	.010	.015	.015	.015	.020	.020	.020	.020
1	.010	.010	.010	.015	.015	.015	.015	.020	.020	.020	.020
1¼	.010	.015	.015	.015	.015	.015	.015	.020	.020	.020	.020
1½	.010	.015	.015	.015	.015	.015	.020	.020	.020	.020	.020
2	.015	.015	.015	.015	.020	.020	.020	.020	.020	.020	.020
2¼	.015	.015	.015	.015	.020	.020	.020	.020	.020	.020	.025
2½	.015	.015	.015	.020	.020	.020	.020	.020	.025	.025	.025
3	.015	.015	.020	.020	.020	.020	.025	.015	.025	.025	.025
3½	.015	.020	.020	.020	.020	.025	.025	.025	.025	.025	.025
4	.020	.020	.020	.020	.025	.025	.025	.025	.025	.030	.030
4½	.020	.020	.020	.025	.025	.025	.025	.025	.030	.030	.030
5	.020	.020	.025	.025	.025	.025	.025	.030	.030	.030	.030
6	.020	.025	.025	.025	.025	.025	.030	.030	.030	.030	.030
7	.025	.025	.025	.025	.030	.030	.030	.030	.030	.030	.030
8	.025	.025	.025	.025	.030	.030	.030	.030	.030	.030	.030
9	.025	.025	.025	.030	.030	.030	.030	.030	.030	.030	.030
10	.025	.025	.030	.030	.030	.030	.030	.030	.030	.030	.030
11	.025	.025	.030	.030	.030	.030	.030	.030	.030	.030	.030
12	.030	.030	.030	.030	.030	.030	.030	.030	.010	.030	.030

SOURCE: The L. S. Starrett Co.

TABLE 21.3 Grinding Wheel Recommendations

Work	Abrasive	Grain	Grade	Structure	Bond	Abrasive	Binding process
Aluminum							
Cylindrical	37	40/3	–K	5		Crystal	Vitrified
Floor stands, 5000–6000 ft/min	37	24	–O	5		Crystal	Vitrified
Floor stands, 7000–9500	12	30 Tr.	–O	4	T	Aluminum	Resinoid
Cutting off, 9000–16,000 ft/min	12	24 Tr.	–S	8	T-2	Aluminum	Resinoid
Brass							
Cylindrical	37	36	–K	5		Crystal	Vitrified
Snagging (floor stands) 5000–6000 ft/min	37 12	24 Tr.	–P	5		Crystal	Vitrified
Snagging (floor stands) 7000–95000 ft/min	37 12	24 Tr.	–P	4	T	Crystal	Resinoid
Cutting off 9000–16,000 ft/min		30	–W	7	T-2	Aluminum	Resinoid
Broaches							
Sharpening	38	46	–K	5	BE	Aluminum	Vetrified
Bronze (soft), use same wheels as for brass							
Bronze (hard)							
Cylindrical		46	–L	5	BE	Aluminum	Vitrified
Snagging (floor stands), 5000–6000 ft/min		20	–O	8	BE	Aluminum	Vitrified
Cutting off 9000–16,000 ft/min		30	–W	7	T-2	Aluminum	Resinoid

	Grain No.	Grit	Grade	Structure	Bond	Abrasive	Process
Bushing (hardened steel)							
Cylindrical		60	−L	5	BE	Aluminum	Vitrified
Internal		60	−K	5	BE	Aluminum	Vitrified
Bushing (cast iron)							
Cylindrical	38	46	−K	5		Crystal	Vitrified
Internal	37	46	−J	5		Crystal	Vitrified
Cast Iron							
Cutting off 12,000–16,000 ft/min	37	36	−R	8	T-2	Crystal	Vitrified
Cylindrical	37	36	−J	5		Crystal	Vitrified
Internal	37	46	−J	5		Crystal	Vitrified
Surfacing (cups and cylinders)	37	16	−H	8		Crystal	Vitrified
Surfacing (straight wheels)	37	30	−J	8		Crystal	Vitrified
Snagging (floor stands), 5000–6000 ft/min	37	20	−R	5		Crystal	Vitrified
Snagging (floor stands), 7000–9500 ft/min	37	16	−P	2	T	Crystal	Resinoid
Snagging (portable grinder), 5000–6000 ft/min	37	24	−S	5		Crystal	Vitrified
Snagging (portable grinder), 7000–8500 ft/min	37	20	−Q	4	T-H	Crystal	Resinoid
Cemented Carbide							
Offhand grinding, cup wheel							
Roughing (dry)	39	60	−I	7		Crystal	Vitrified
Roughing (wet)		100	−R	25		Diamond	Resinoid
Finishing (dry)	39	120	−H	7		Crystal	Vitrified
Finishing (wet)		220	−R	50		Diamond	Resinoid
Cutter, grinding							
Backing-off, cup wheel							
Roughing (dry)	39	60	−I	7		Crystal	Vitrified
		100	−R	100		Diamond	Resinoid
Finishing (dry)	39	100	−H	7		Crystal	Vitrified
		220	−R	100		Diamond	Resinoid

5. Rotary table, using a cup or cylinder wheel mounted on a vertical spindle.

All of the above machines possess advantages for certain classes of work. The amount and variety of the work ground, as well as the character of the surface required, determine the type of machine to select.

SELECTED CHARTS AND TABLES USED IN GRINDING APPLICATIONS

The data in Table 21.1 give grinding wheel speeds for various applications while Table 21.2 gives allowances for grinding. Tables 21.3 contains typical grinding wheel recommendations as furnished by the Norton Company.

22

Laps and Lapping

Merely grinding metal—as discussed in Chap. 21—at its best is relatively rough when subjected to magnification. However, these imperfectins can be removed by lapping, which requires a good lapping tool (lap), a compound suitable for the job, and some patience. The compound should be approximately the same hardness as the material to be lapped, while the lap itself should be considerably softer than the material. A single lapping compound will not serve all purposes. Therefore, manufacturers of lapping compounds provide guides for selecting the proper compound for a particular lapping job.

LAPPING COMPOUNDS

Both external and internal universal laps are available on the market, but most shops, because of varied applications, make their own. In general, for an external lap, start with the object to be lapped, apply the appropriate lapping compound generously to it and also to the lap, and process either by hand polishing or else have the object rotating in a lathe and then apply the lap by hand. The revolving speed should be approximately 300 rpm and the lap should traverse the work slowly and with light pressure. Check the work frequently for size and finish, and use a hot, strong cleaning solution to clean all surfaces before finally drying them with a clean soft rag.

The lap may be very simple, consisting of a strip of abrasive cloth attached to a shaft, or it may be elaborately constructed of lead, copper, cast iron, and the like. A cast iron lap used for lapping the inside of a hardened steel bushing is shown in Fig. 22.1. The workpiece may be secured in the lathe chuck, while the lap, on a tapered mandrel, is secured in a drill chuck in the tail stock of the lathe. While the work rotates, the lap is slowly fed into the bushing. Conventional valve lapping compound will suffice for most applications, but Clover abrasive compound kits seem to suit most needs of the average metalworking shop. Two kits (no. 4 and no. 6) seem to be the most popular.

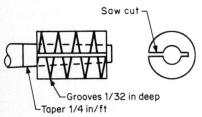

Saw cut

Grooves 1/32 in deep

Taper 1/4 in/ft

FIG. 22.1 A cast-iron lap used for lapping the inside of a hardened steel bushing.

Clover No. 6 Kit

Six different grit sizes are available in this kit. The grit consists of silicon carbide formulated with a special grease base that lasts a long time and does not break down under heavy use. The sizes (120 and 800 grit) are ideal for nearly all fitting, lapping, and running-in operations. They are also good for the initial work or reworking of metal-to-metal fits where both metals are steel.

The abrasives in kit no. 6 are not designed for use between steel and aluminum alloys, as silicon carbide crystals will embed in the softer metal. The kit contains 2-oz metal cans in the following grits:

- E, 120 (coarse)
- D, 180 (medium coarse)
- B, 240 (med fine)
- 1A, 320 (very fine)
- 4A, 600 (extremely fine)
- 5A, 800 (microfine)

Clover Kit No. 4

This kit contains four different grit sizes of ultra fine water-soluble aluminum oxides, and is formulated with a petroleum-base emulsion-type carrier which is water-soluble and easily removed from the work surface once finished. It is particularly useful for lapping inside surfaces where solvents may be difficult to apply. The crystal used is aluminum oxide, which has the characteristics of being rounded like river rock and breaks down into smaller and smaller rounded pebbles as it is used. These crystals are extremely tough, have great durability and wear resistance, but they do not embed because they are not sharp. While these compounds cut slower than others they are recommended for metal-to-metal fitting situations where steel is fitted to aluminum alloy, as the crystal will not embed into the softer metal. The grit sizes in this kit are as follows:

- 4A, 600 (14 microns)
- 5A, 800 (9 microns)
- 6A, 1000 (5 microns)
- 7A, 1200 (3 microns)

Three types of lapping medium are normally used on lapping machines in the industry:

1. Metal laps and loose abrasive mixed with a lubricant.
2. Bonded abrasives for commercial production work.
3. Abrasive paper or cloth.

One common form of lap consists of a split sleeve and a tapering holder or arbor which is used to expand the sleeve to fit the hole being lapped. Lapping compound is applied to the lap and then moved gradually through the hole until the desired finish is obtained. The split in the lap allows it to be expanded as it becomes worn.

The bushing forming the lap frequently is made of cast iron or copper. The actual lapping operation consists in traversing the lap in and out so as to cover the entire surface of the hole in the case of internal lapping. The lap length should be about 3 times greater than the diameter of the hole being lapped.

Care should be exercised when applying the abrasive compound, especially toward the final stages of the operation, for if too much is applied, the ends of the hold will have a tendency to enlarge because of the abrasive particles crowding under the edges of the lap.

Laps used for smoothing flat surfaces are usually made of cast iron. To obtain accurate results with this type of lap, the lapping surfaces must be exceptionally true. Therefore, a flat lap must be carefully planed and then should be carefully scraped, or otherwise trued, to a standard surface plate.

To charge a flat lap, spread a very thin coating of the prepared lapping compound over the surface and press the small cutting particles into the lap with a hard steel block. There should be the minimal amount of rubbing involved in this operation. When the entire surface seems to be charged, clean and carefully examine for any bright spots. If any are found, keep charging the surface until a uniform gray appearance is obtained.

After the lap is charged, all loose abrasive should be cleaned off with an appropriate cleaner such as AWA 1,1,1. During the lapping process, it is recommended that the surface be kept moist with kerosene or other recommended solution. Other liquids may cut faster, but they have a tendency to evaporate quickly and must be reapplied often. If a lap is correctly charged from the beginning, and kept moist, it will carry all the abrasive that is required for the job for a long time. The pressure applied to the surface being lapped will vary, but in general it should be only enough to make good contact. The lap will only cut so fast, and any attempt to increase the speed will result in the lap becoming stripped in places, requiring that it be recharged before it will function properly again.

A rotating disc is sometimes used for lapping flat surfaces. The lap is usually made of cast iron. The rotating disc is attached to an arbor and the lap itself should be turned on the arbor on which it is to be used, as it is almost impossible to put a lap back on an arbor accurately after it has been removed.

Rotary-disc laps are also made of soft steel with their sides relieved so that there is only a narrow lapping surface in comparison to the entire disc. Their most practical use is for lapping sharp corners on hardened work where no other means is practical.

One consideration that is often overlooked in lapping is the temperature of the lap and workpiece. With an accurate plane surface lap, it is possible to produce surfaces that are either plane, convex, or concave, depending upon the temperature of the lap in relation to the work. If the work is warmer than the lap, the result will be a convex surface; colder work will result in a concave surface, and equal temperature will result in a plane surface.

LAPPING OPERATIONS

In general, lapping is a rubbing process for removing minute amounts of metal from surfaces that must be precision-finished in order to:

1. Produce true surfaces
2. Improve dimensional accuracy
3. Correct minor surface imperfections
4. Provide a close fit between two surfaces

When lapping is done precisely, metal can be brought to a given dimension with greater accuracy and a finer finish than by any other means.

Lapping operations fall into two categories: hand lapping and machine lapping. Hand lapping is frequently used in the average machine shop, while machine lapping is utilized most frequently in industrial applications where speed is of the utmost importance. Hand lapping, although slow and tedious, avoids the high cost of lapping machines, and every machinist should know the basic principles of hand lapping for trueing external, internal, and flat surface. Obviously, hand lapping is employed for lapping small quantities of work.

Machine Lapping

Machine lapping employs mechanical motions to control the movement of the work in relation to the lap, and also to automatically adjust the pressure of the lap upon the work. Common lapping machines are available for external cylindrical, flat, spherical, and gear lapping on a production basis.

Laps for these machines are generally made of some material soft enough that the abrasive can be readily pressed into the surface; the process is known as "charging." Soft close-grained cast iron, copper, brass, or lead may be used for the lap. However, cast iron is usually the preferred material; soft steel is the least common.

Vertical lapping machines with cast iron laps and loose abrasive are used for both cylindrical and flat workpieces. The universal lapping machine can also be used for both types of work. What is called a "centerless" lapping machine can be used to improve the accuracy and finish of commercially ground parts. When used, finishes can be obtained within 1 to 2 microinches; that is, diameter tolerances within .00005 to .000025 in are possible on the centerless lapping machine. These parts may also be lapped at a rapid pace. For example, when used in conjunction with a centerless grinder, the centerless lapping machine can keep pace with the production rate of the grinder if the lapping machine is placed at the end of the production line.

PRACTICAL APPLICATIONS

Cylindrical objects, such as internal-combustion-engine cylinders, that are only slightly out of round or slightly tapered can be trued up by lapping. An untrue cylinder is positioned in the appropriate lapping machine, the lap charged, and the lapping process begun. During the operation, measurements are checked with either an inside micrometer or a dial-gage indicator. When the gages show that the cylinder is true for roundness and lack of taper, the operation is stopped, the cylinder is cleaned with AWA 1,1,1 or similar cleaning agent, and another out-of-true cylinder is placed in the machine for lapping.

LAPPING RIFLE BARRELS

Any machinist working in the firearms industry knows that the bore dimensions and smoothness of the rifling in a rifle barrel (see cross section shown in Fig. 22.2) is of the utmost important to obtain the greatest accuracy—the ultimate purpose of a modern rifle. Most barrel blanks are currently being rifled by the broaching method, and although some very smooth work is obtained, lapping always improves them. Lapping is also an excellent way to restore older, worn barrels to acceptable accuracy.

A rough, pitted bore tends to pick up leading or metal fouling as projectiles are fired through the bore. Cleaning such a bore is

Proper crown   Improper crown

FIG. 22.2 Cross-section of a rifle barrel.

also a chore. In all cases, such a barrel will be less accurate than a smooth, well-polished bore.

Since rifle barrels have a rifled interior, general polishing—as discussed previously—will not work, since the rifling will be removed in the process. In this case, the lap must spin with the rifling in the bore, as it is being pushed up and down the interior of the barrel. The first step, therefore, is to remove the barrel from the receiver, and take accurate interior measurements of the lands and grooves. Calipers may be used for taking measurements at each end of the barrel, but for best results, use a soft lead slug slightly larger than the bore itself. This lead slug can then be started into the muzzle end of the barrel, and then pushed or gently tapped through the entire length of the barrel to fall out the breech end. In doing so, the lead slug will give a perfect imprint of the rifling (lands and grooves) on the inside of the barrel, and dimensions can then be obtained by measuring these imprints with a micrometer. Make a note of the findings.

Continue by making a lapping rod out of steel with the diameter as large as will slide easily through the bore. The head of this rod should be turned or filed as shown in Fig. 22.3. The opposite end of the head should then be brazed, silver-soldered, or welded to a handle assembly so the rod will freely turn on a smooth working ball bearing. Several methods of obtaining a free-turning rod are possible. Belding & Mull of P.O. Box 428, 100 N. 4th St., Phil-

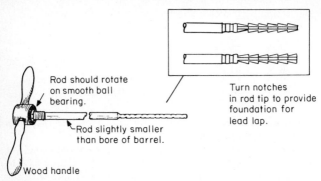

Rod should rotate on smooth ball bearing.

Rod slightly smaller than bore of barrel.

Wood handle

Turn notches in rod tip to provide foundation for lead lap.

FIG. 22.3 Details of a barrel lapping rod.

lipsburg, PA 16866 manufacture stainless steel cleaning rods with ball bearing handles that work very well as lapping rods. An old bicycle pedal may be used; merely braze the lapping rod to the threaded spindle end and then attach a hardwood handle. Or a machine shop may design its own ball-bearing handle for use on the lapping rod.

Once the rod has been made, secure the barrel in a vertical position in a bench vise. Wrap a piece of cotton twine around the lapping rod just below the rod tip, enough to provide a snug fit inside the barrel. Now insert the rod from the breech end of the barrel, pushing it in and stopping it about ½ in from the muzzle of the barrel. Heat the barrel with a propane or other relatively low-heat torch until the barrel is just about hot enough to melt solder, yet not hot enough to affect the metal finish. When this temperature is reached, use a ladle and carefully pour melted lead into the bore to form a solid slug around the tip of the rod, as shown in Fig. 22.4. Let this slug cool slightly, and then push it about an inch or

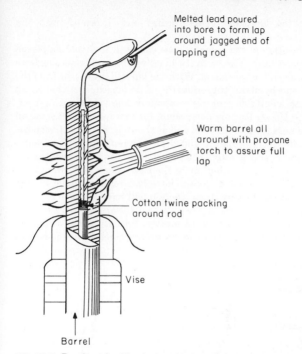

Melted lead poured
into bore to form lap
around jagged end of
lapping rod

Warm barrel all
around with propane
torch to assure full
lap

Cotton twine packing
around rod

Vise

Barrel

FIG. 22.4 Pouring a lead lap into a rifle barrel.

so out of the muzzle of the barrel. Be careful not to push it out all the way, or the job will have to begin all over again. Note that the slug now has the full imprint of the barrel's rifling and is molded to the exact dimensions of the bore. With the lap partially protruding from the muzzle, file off any burrs or fins that may prevent the lap from being pushed and drawn through the barrel.

The position of the barrel is now changed. It should be placed horizontally in the vise, with the handle of the rod projecting from the breech end of the barrel. With the lap still projecting from the muzzle end by about three-fourths of its length, coat the lap all over with a suitable lapping compound: fine valve lapping compound or Clover lapping compound, for example. Squirt some oil into the bore from the breech end as shown in Fig. 22.5. Grasp the rod handle in both hands and draw the lapping rod backward, slowly and evenly through the barrel, until the about half the lap projects into the chamber, or in the case of no chamber, out the breech end of the barrel. Again, be very careful not to let the lap project out all the way; if so, the job will have to begin all over. Now push the rod so the lap travels back toward the muzzle. Keep up this backward and forward motion. Recoat the lap with compound as often as necessary to maintain a smooth cutting action.

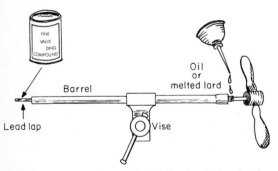

FIG. 22.5 Position barrel should be in during lapping operation.

The amount of time required to obtain a good job depends on the condition of the bore prior to beginning the lap. Some jobs have been completed in 20 to 30 strokes (back and forth motions) while others have taken hours to complete. During the lapping operation, if the lap becomes worn or loose, it will have to be melted down and another lap cast.

Often, lapping a rifle barrel will remove every pit and tool mark in a very short period of time, resulting in a fine, smooth, accurate barrel comparable to the very best obtainable. However, even if every mark is not eliminated by lapping, an improvement always results from the job.

When the operation has been completed, withdraw the lapping rod completely. Clean the bore of the barrel with AWA 1,1,1 or similar solvent, using a tight-fitting cleaning patch on a rifle cleaning rod. A better cleaning will result if the entire barrel is soaked in a solution of Brownell's d'SOLVE, then blown dry with compressed air. Then the bore is oiled with a recommended gun oil placed on a cleaning patch and pushed through the bore with a cleaning rod.

23

Punches and Shears

PUNCHES

A punch is a blunt tool used to pierce a solid material. Two types are in normal use: hand and machine. Unlike shears, which actually cut the metal, a punch pushes or tears the metal, and each punch is specifically designed for forming or punching a particular part.

A cold chisel is another tool related to the punch, but the chisel actually cuts or chips. A variety of cold chisels are used in the machine shop, but the most common is the cold chisel which is designed specifically for cutting metal. Its use is varied, but the removal of rusted rivets and nut heads is probably the most common usage of the cold chisel in the machine shop.

The diamond-point chisel is used for chipping v-shaped oil grooves and sharp corners, while the cape chisel is forged to produce a cape or flare for the widest flat at the cutting edge. The cape chisel is used for cutting narrow slots, keyways, and rectangular grooves.

The round-nose chisel is used for producing oil grooves and other concave surfaces. This chisel is used for drawing a drill back to the true location of the hole to be drilled.

The round-handled solid punch shown in Fig. 23.1 and the one-piece hollow punch in Fig. 23.2 are both commonly used in machine-shop applications—especially on sheet metal. Both are tapped with a hammer. Other types of hand punches use interchangeable bits or punches in a device resembling pliers, and various sizes of punches and dies can be used to produce the desired sizes and shapes of the openings.

For larger work, machine punches are used, and various types of punches are available for practically any desired application. The parts of one machine punch include the plunger, coupling, stripper, punch, die and socket, to name just a few. In use, the punch is secured to the plunger by the coupling nut which bears against a shoulder on the punch. the die fits into a socket, and is held in position by setscrews.

Great force is required to drive a punch through a piece of heavy metal, and therefore, most of the time, a heavy-rimmed flywheel is used to distribute the work of the machine over the entire period of revolution of the drive shaft. Belt power is the normal

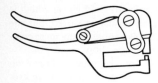

FIG. 23.1 Round-handle solid sheet metal punch.

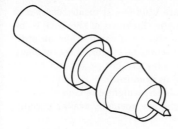

FIG. 23.2 One-piece hollow punch.

medium used to accelerate the speed of the flywheel during the greater part of the revolution of the drive shaft. The energy stored in the flywheel is expended at the expense of its velocity during the portion of the revolution it does work. Some typical punch designs are shown in Fig. 23.3.

SHEARS AND SNIPS

Hand snips resemble an ordinary pair of scissors, in that two handles are used, with cutting edges on the ends of each, to move backward and forward; as the cutting edges move together, the material is cut. Straight snips are used to cut straight lines and

FIG. 23.3 Two typical punch designs.

outside curves; hawksbill snips are used to cut inside curves and intricate work. Depending upon the thickness of the work to be cut, the handles on snips may be about the same size as an ordinary pair of scissors, or the handles may be longer. Sheets of sheet metal too thick to be cut with snips are cut with hand shears which have handles as long as 18 in or more.

Aviation snips are used to cut both straight and curved cuts in thin-gage sheet metal. Three styles are normally available: right, left, and straight cut.

Bench shears are found in almost every sheet-metal shop, and machine shops with much sheet metal to cut will also find this machine useful. The hand- or foot-operated shears can be used to cut sheets of metal that are too large to be cut with hand snips. Motor-driven shears are available in many types and are also used to cut sheet metal.

In addition to the cutting and punching operations that are performed with snips, shears, and punches, metal can also be shaped into arcs, spirals, circles, cylinders, and cones by means of forming equipment. These machines vary from small hand-operated forming rolls to large motor-driven bending rolls and forming presses.

The standard hand-operated forming rolls are 30 in in width, and are used to bend sheet metal to a curved form. The distance between the geared rolls can be regulated with adjusting screws by raising or lowering the lower front roll, depending on the sheet metal, to a smaller radius, or it can be lowered to form a cylinder with a larger radius.

Hand-operated sheet-metal brakes are widely used in shops. Mechanical and hydraulic press brakes are in the use where large pieces of sheet metal are to be handled. Even though tonnage is sometimes used to express the capacity of a brake, it is the flywheel energy that is required to exert the pressure in the working stroke. The capacity of the brake required to form a single bend in a given material depends on the length of bend, sharpness of bend, and thickness of material.

In modern sheet-metal shops doing a substantial business, computer-operated sheet-metal forming machines can cut, bend, form, and join sheet metal in almost any conceivable form. However, for the average machine shop, probably having very little sheet-metal work, the hand brake will usually suffice. This brake, combined with hand snips and shears can accomplish a surprising amount of work in the hands of capable machinists.

24

Broaching

Broach is the name given a metal-cutting tool especially adapted to finishing various boring operations such as the rifling in a rifle barrel, internal keyways, and many others. The cutting tool used in broaching is straight, with a series of cutting teeth that gradually increase (usually in depth) in size. One or more broaches may be needed to complete a given operation, depending on the extent of the finishing required.

In practice, considerable power—such as that from a hydraulic press—is used to push or pull the broach through a prebored hole. With each push or pull, a small amount of metal is removed from the workpiece by each tooth along the entire length of the run. For this reason, the beginning teeth on a broach are very shallow, removing only a very minute amount of metal. The intermediate teeth usually remove the most metal from the cut, while the last teeth finish the cut to the proper size. In some cases, more than one broach is used to make the cut.

Various types of broaches are shown in Fig. 24.1. Note that the

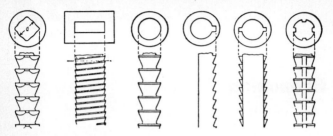

FIG. 24.1 A few types and shapes of broaches currently in use.

face angle of the teeth vary on different broaches. This angle is of the utmost importance, as it allows the teeth to cut properly. Most angles will be from 0 to 20 degrees. The distance, or span, between the teeth is the pitch. The land must be of sufficient strength to withstand the cutting strain, which is tremendous when compared with other metal-cutting tools.

As can be seen in the illustration, a broach may take a variety of forms. Basically a broach cuts metal as it is being forced through or past it, the action being much like that of a metal file. The broach is used for forming square, triangular, and other irregular-shaped holes; for cutting splines and keyways, both internal and external; for finishing bored holes to close tolerances; for burnishing to give a smooth finish; and for surfacing or grooving various parts. In general, a broach usually holds its cutting accuracy and sharpness longer than most metal-cutting tools, and can be designed so that numerous resharpenings do not destroy those qualities.

The average machine shop can find numerous ways of applying broaching to solve many difficult cutting problems, and also to limited production operations. While broaching in large manufacturing plants is done with elaborate machines, the smaller machine shops need little more than a quantity of tool steel for making the broaches, and some method of forcing them into, through, or past the work. For the smallest applications, often a hefty bench vise will suffice; and some broaches even may be hammered through the work, but the most conventional method is to use an arbor press.

The very basic broach will consist of a square-ended tool whose cross section is the size and shape of the hole to be punched. Such broaches may be made from old chisels or from drill rods or other tool steels suitably hardened.

To make multiple-tooth broaches, use high-speed steel, carbon tool steel, carbon-vanadium steel, or any other type of tool steel that can be hardened and will maintain its hardness.

In one instance, the broach blank is machined between centers to exact length and an average of about ¹⁄₁₆ in oversize in diameter. The tail stock on the lathe may then be set over and the broach blank is then tapered. With a formed tool, cut the tooth reliefs on the blank, leaving about ¹⁄₃₂ in oversize for later grinding. The slots or inverted splines on the blank are then milled. To harden, heat slowly to 1500°F, then rapidly to 2350°F; cool rapidly in an air blast to about 1550°F, then allow to cool slowly to room temperature. This procedure, or course, is for one type of tool steel; each type will require a slightly different technique of hardening.

Once hardened, the broaches are once again mounted between lathe centers, with the tail stock set over for tapering. Then a tool-post grinder is used to grind the outside diameter to within the required limits (usually within .002 in of the finish size). At this point the compound rest is swung 3 degrees and the relief angle cut on each tooth, and each tooth is checked for the required size.

The compound rest is again set 10 degrees to the left, to allow the grinding wheel to grind the front rake and fillets at the base of the teeth. Serrations can be formed with a surface grinder.

To use a broach to, say, finish splined holes in a quantity of small gears, the gear blank is laid in a jig recess made especially for this application, a rough hole is then ground, and the small end of the required miltiple-tooth braoch is inserted into the hole; the other end of the broach is set in an arbor press ram hole. The ram is brought down, forcing the broach through the work and into a receptacle beneath the jig. This brings the hole in the gear halfway to shape. Then another broach is used, and pushed through the work in the same manner, which completes the job. In a case like this, two broaches are usually sufficient to enlarge a hole .1060 in. For other jobs, a dozen or more may be required.

Broaches will cut cleaner and easier and last longer if a cutting fluid is used with them. For steel and iron, a lard oil or a sulfurized cutting oil may be used. Bronze and brass usually require no lubricant. Kerosene or a special cutting fluid may be used with aluminum.

Burnishing broaches have teeth that do no cutting but merely push or squeeze the metal into smoothness and exact size. Often the final teeth or diameter of a forming broach burnish the metal surface to complete smoothness. Usually the final half-dozen or so teeth on a multitooth broach are all the same size. These act to smooth the hole; but there is another reason for them: as a broach is resharpened, the starting teeth are reduced in size. Eventually no. 2 tooth will be the same size that no. 1 tooth was originally; no. 3 will become no. 2, and so on. When the last cutting tooth is reached, it will have become the size of the next-to-the last one. If there were no additional teeth, the finished dimension would be off. But by having several extra, full-sized teeth, there is margin for a great many resharpenings before the very last tooth has to be reduced in size. That is one reason why a broach, although costly to make in most cases, lasts longer and produces more accurate work than any other kind of cutting tool.

Multitooth broaches may be either pushed or pulled through the work. The push type has to be made comparatively short and stubby, for strength; and the length seldom is more than a foot or so. By using a series of short broaches, the same cutting capacity is obtained with a single lone one, but the time required for a complete cut is greater. For the average machine shop, push-type broaches are easier to operate with equipment usually at hand. The pitch or spacing of broach teeth should be such that at least two teeth are always in contact with the work. Frequently the broaching action holds the work firmly, and no special jigs or fixtures are required.

25

Metal Spinning

SPINNING OPERATION

Metal spinning, basically, consists of spinning a flat disc of metal at high speed; then it is pressed and shaped by means of hand tools. Metal forming dies can be built to do the same thing, but dies are expensive and spinning is the preferred method when only a few articles are to be formed.

The essential parts of a metal-spinning operation are shown in Fig. 25.1, and while this setup resembles a conventional metal-turning lathe, metal spinning requires different techniques—and a different degree of expertise and skill. Good spinning techniques result in a finished piece with variations in thickness of less than 25 percent of the original thickness. Faulty spinning results in a finished product that appears to be stretched or has great variation in thickness.

Materials such as iron, soft steel, pewter, copper, brass, zinc, aluminum, silver, and britannia are all suitable for spinning, but

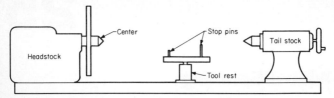

FIG. 25.1 Essential parts of a metal-spinning operation.

the thickness and type of metal will determine the rotation speed. For example, zinc is easy to work and may be rotated from approximately 1000 to 1400 rpm. On the other hand, iron or soft steel is the hardest and must be rotated not faster than 600 rpm; other metals will be rotated somewhere between these speeds.

SPINNING TOOLS

A large variety of spinning tools are required in metal spinning. All are made from tool steel forged to the required shape and size. Most spinning tools are relatively long (about 3 ft) to obtain the required leverage to exert great pressure on the material being worked. In use, the wooden handles of the tools are normally held under the armpit of the worker, so that sufficient physical power can be exerted to form the desired shape.

Several shapes of metal spinning tools are shown in Fig. 25.2. Of these, the ball-and-point tool is the one most frequently used. It is used to start the work and bring it to the desired rough shape. The smoothing tool is then used to obtain the desired finish on the object. The few examples shown are but a sampling of those available and used in metal spinning operations.

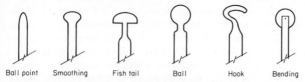

| Ball point | Smoothing | Fish tail | Ball | Hook | Bending |

FIG. 25.2 Several shapes of metal-spinning tools.

Templates, called "chucks," are used as molds to shape the object. Again, like the forming tools, chucks come in a variety of shapes and sizes. In general, the shape of the chuck will conform to the desired shape of the finished article. Solid chucks are used for objects permitting the withdrawal of the chuck after the object has been formed. Sectional chucks are those designed to be dismantled after the forming has been completed to allow the withdrawal of the chuck. In some cases, one section of the chuck is removed at a time as a certain portion of the object is completed.

Traditionally, spinning chucks were made of kiln-dried maple, which permitted them to be turned on a wood-turning lathe to the required dimensions. However, in recent years, spinning chucks have been made from cast iron or steel.

Another essential item for metal spinning is the follower, which consists of a block of wood and is used to hold a sheet-metal blank against the chuck or form for metal spinning. Again, several types and shapes are in normal use.

Spinning metal, to a certain degree, is more of an art then merely a mechanical operation. Much practice is required to attain a high degree of skill. Even the fundamentals are rather complicated, and require much more space that can be allotted here. However, a brief explanation of a metal-spinning operation will serve to give a general idea of how the technique is accomplished.

As with most machine-shop operations, the mounting of the work is very important. First the chuck is turned onto the spindle, after which the blank (the piece that will be formed) is pressed against the chuck by hand. Then the tail stock center is turned against the center of the follower with just enough pressure to hold the blank in place. To produce an accurate finished object, the blank must not be centered perfectly. While there are several modern methods that may be used for this operation, the simplest is to use a hardwood stick pressed against the edge of the blank after starting the lathe at a very slow speed. Once the blank is centered, more pressure is applied to the center to securely clamp it in place. This mounting is very important for safety reasons. If improperly mounted, or if mounting is attempted while the lathe is in motion, the sharp piece of metal can spin outward and away from the lathe, acting as a free-flying buzz saw through the air, possibly severing arms, necks, or any other thing it may come into contact with. So much caution must be used at this stage of the operation.

A tool rest is provided to allow the operator to obtain the desired pressure for forming the finished object. The prop pin on the tool rest should be adjusted to a point near the blank, so that adequate leverage can be obtained. The operator then holds the handle of the tool under the right (or left) armpit, using the spinning tool as a lever and the prop pin on the tool rest as a fulcrum. As pressure is applied, the metal disk is forced slowly against the chuck, and gradually takes on the rough shape of the chuck.

While the spinning tools are being used, they should be kept moving constantly, back and forth from center to edge, from edge to center, etc., and pressure should be applied only when the tool is moving toward the edge of the blank; never while moving toward the center. At the same time, the back stick, held in the left (or right) hand, should be pressed firmly against the opposite side of the metal at a constantly changing point opposite the tool. This prevents wrinkles in the work.

Depending upon the final shape of the finished object, several different shapes of followers will be used to obtain the final form. In other words, the shaping is done in steps, using different tools and followers until the required shape is obtained, and with the desired finish.

During the spinning process, a gradual hardening can occur. This hardening must be corrected before further spinning may be done as the blank is sure to crack otherwise. When hardening occurs, the metal blank must be removed from the lathe and annealed before continuing. The method of annealing will differ with different types of metal, but in all cases, the metal should be fluted or dented by hammering with a rawhide hammer to relieve the stresses and to prevent cracking during the annealing process.

After annealing, the oxide scale formed in the annealing process should be removed with a pickling solution. Then the work should be washed thoroughly in running water and permitted to dry. In practice, there are techniques that can prevent scaling during the annealing process. One is to wrap the object in stainless steel "paper" during the heating process.

Due to the hand work involved in this type of machine-shop application, metal spinning is rarely used when articles are produced in large quantities; forming dies are used instead. However, when a single article is needed, and the equipment is available in the shop, spinning is one relatively inexpensive way to obtain the desired article. For example, the follower to an obsolete tubular magazine rifle can easily be made by metal spinning. The chuck and followers can be turned out of wood on the lathe, and the exact form may be produced by spinning of thin sheet metal.

26

Gears

In general, gears are used to transmit positive motion and power from one revolving shaft to another. Most gears take the form of a wheel, but some are in the form of a shaft, cone, or other shape. Also, most gears have teeth which mesh with the teeth of other gears to provide a positive-motion drive, so that when one of the engaging gears turns, so will the other.

GEAR NOMENCLATURE*

Addendum a.   The height by which a tooth projects beyond the pitch circle or pitch line.

Base diameter D_b.   The diameter of the base cylinder from which the involute portion of a tooth profile is generated.

* SOURCE: Boston Gear Co.

Backlash B.   The amount by which the width of a tooth space exceeds the thickness of the engaging tooth on the pitch circles. As actually indicated by measuring devices, backlash may be determined variously in the transverse, normal, or axial planes, and either in the direction of the pitch circles or on the line of action. Such measurements should be corrected to corresponding values on transverse pitch circles for general comparisons.

Bore length.   The total lenth through a gear, sprocket, or coupling bore.

Circular pitch p.   The distance along the pitch circle or pitch line between corresponding profiles of adjacent teeth.

Circular thickness t.   The length of arc between the two sides of a gear tooth on the pitch circle, unless otherwise specified.

Contact ratio m_c.   In general, the number of angular pitches through which a tooth surface rotates from the beginning to the end of contact.

Dedendum b.   The depth of a tooth space below the pitch line. It is normally greater than the addendum of the mating gear to provide clearance.

Diametral pitch P.   The ratio of the number of teeth to the pitch diameter.

Face width F.   The length of the teeth in an axial plane.

Fillet radius r_f.   The radius of the fillet curve at the base of the gear tooth.

Full-depth teeth.   Those in which the working depth equals 2.000 divided by the normal diametral pitch.

Gear.   A machine part with gear teeth. When two gears run together, the one with the larger number of teeth is called the gear.

Helix angle ψ.   The angle between any helix and an element of its cylinder. In helical gears and worms, it is at the pitch diameter unless otherwise specified.

Hub diameter.   The outside diameter of a gear, sprocket, or coupling hub.

Hub projection.   The distance the hub extends beyond the gear face.

Involute teeth.   In spur gears, helical gears, and worms, those in which the active portion of the profile in the transverse plane is the involute of a circle.

Keyway.   The machined groove running the length of the bore. A similar groove is machined in the shaft and a key fits into this opening.

Lead ψ.   The axial advance of a helix for one complete turn, as in the threads of cylindrical worms and teeth of helical gears.

Long- and short-addendum teeth.   Those of engaging gears (on a standard-designed center distance), one of which has a long addendum and the other has a short addendum.

Normal circular pitch p_n.   In cylindrical gears, the distance along the normal helix on a pitch cylinder between corresponding profiles of adjacent, equally spaced teeth.

Normal circular thickness t_n.   The circular thickness in the normal plane. In helical gears, it may be considered as the length of arc along a normal helix.

Normal diametral pitch P_n.   The value of the diametral pitch as calculated in the normal plane of a helical gear or worm.

Normal plane.   The plane normal to the tooth surface at a pitch point and perpendicular to the pitch plane. For a helical gear this plane can be normal to one tooth at a point lying in the plane surface. At such a point, the normal plane contains the line normal to the tooth surface and this is normal to the pitch circle.

Normal pressure angle ϕn.   The pressure in a normal plane of a helical tooth.

Operating clearance c.   The amount by which the dedendum in a given gear exceeds the addendum of its mating gear.

Operating pressure angle ϕ_r.   Determined by the center distance at which the gears operate. It is the pressure angle at the operating pitch diameter.

Outside diameter D_o.   The diameter of the addendum (outside) circle.

Pinion.   A machine part with gear teeth. When two gears run together, the one with the smaller number of teeth is called the pinion.

Pitch circle.   The circle derived from a number of teeth and a specified diametral or circular pitch. The circle on which spacing or tooth profiles is established and from which the tooth proportions are constructed.

Pitch cylinder.   The cylinder of diameter equal to the pitch circle.

Pitch diameter D.   The diameter of the pitch circle. In parallel shaft gears, the pitch diameters can be determined directly from the center distance and the number of teeth.

Pressure angle ϕ.   The angle between a normal to the tooth profile in that plane and the line of intersection of that plane with the corresponding pitch planes. In involute teeth, pressure angle is often described also as the angle between the line of action and the line tangent to the pitch circle. Standard pressure angles are established in connection with standard gear-tooth proportions.

Tip relief.   An arbitrary modification of a tooth profile whereby a small amount of material is removed near the tip of the gear tooth.

Undercut. A condition in generated gear teeth when any part of the fillet curve lies inside a line drawn tangent to the working profile at its point of juncture with the fillet.

Whole depth h_t. The total depth of a tooth space, equal to addendum plus dedendum, equal to the working depth plus variance.

Working depth h_k. The depth of engagement of two gears, that is, the sum of their addendums.

See Fig. 26.1 for gear tooth parts.

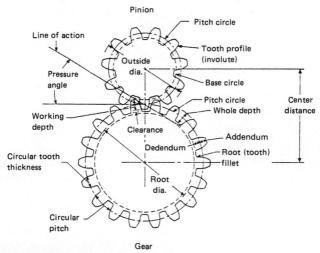

FIG. 26.1 Gear tooth parts.

27

Threads and Threading Operations

Threads in metal work may be produced by several different methods: grinding, rolling, milling, or using a screw-cutting tap or die, but a metal-turning lathe is the preferred method when the threads must be concentric with other turned surfaces. Threads of various pitches may be cut on the lathe by connecting the head stock spindle of the lathe with the lead screw via a series of gears so that a positive carriage feed is obtained and the lead screw is driven at the required speed in relation to the head stock spindle.

The lead screw on one popular lathe, for example, has 8 threads per inch and the gearing between the head stock spindle and the lead screw may be arranged so that practically any desired pitch of the threads may be cut. To illustrate, if the gears are arranged so that the head stock spindle revolves four times while the lead screw revolves once, the thread cut will be 4 times as fine as the thread on the lead screw or $4 \times 8 = 32$ threads per inch. Figure 27.1 shows a lead screw dial indicator.

Either right-hand or left-hand threads may be cut by reversing

FIG. 27.1 Lead screw dial indicator. *(South Bend Lathe Co.)*

the direction of rotation of the lead screw. This may be accomplished by shifting the tumbler reverse lever on most lathes.

Threads may also be external or internal. An external thread is a thread on the outside of a member, as on a threaded pipe. An internal thread is a thread on the inside of a member, such as a pipe coupling.

To cut an external thread, the thread-cutting tool is fed into the work by means of a compound rest feed. This rest is sometimes set to advance the tool squarely into the work, but many shops prefer to set the feed on a 30 degree angle; the cutting tool then removes metal only with its left-hand edge and tends to produce a smoother thread than if it were fed straight in. When the tool moves into the work at a 30 degree angle it is advanced 1.154 times the actual depth of the thread.

The cross slide is used in backing off a tool after a cut has been completed so that the carriage can be returned for another cut.

The cross slide is then moved into its original position, located either by a stop or by a reading on the cross-slide index collar. To prevent the tool from tearing the threads, the lathe spindle speed should be slow, say around 28 rpm to begin with. Once experience has been gained on the size and type of metal being threaded, this speed may be increased somewhat. The gears which drive the lead screw must be arranged to produce the specified number of threads per inch.

Once the first shallow trial cut is made, every cut thereafter must track in the same spiral, which requires that the lathe carriage be engaged to the lead screw at exactly the right moment. This is easily done with a threading dial. If an even number of threads per inch is being cut, the lead screw can be engaged when any mark on the dial comes opposite the index mark. In cutting an odd number of threads per inch, the lead screw is engaged at a numbered mark on the dial.

THREAD FORMS

In modern industry there are many different types of thread forms. Those in normal use are described below.

The V Thread

The American Society of Mechanical Engineers (ASME)—both fine and coarse—are practically standard for machine shop work in the United States. These threads are 60 degree, v threads. Other types of v threads include the British Standard Whitworth, British Standard Fin and British Association (B.A.). Actually, few of these are true v threads; most of them have their points cut off so that the depth of them is 75 percent of the depth of a true v thread of the same pitch.

The cutting tool is usually ground for cutting sharp-pointed v threads and the thread is cut with the regulation v bottom, but the top is left with the proper amount of flat. Only when the utmost strength is needed should the tool be ground to the exact National Form. The tool, however, may be ground to cut an exact National Form v thread by flattening the sharp point so that it will fit in the selected slot of the ASME gage. The general sharp-pointed 60 degree v thread screw will work satisfactorily with the corresponding nut.

The Acme Screw Thread

This type of thread is often found in power transmissions, where heavy loads require close-fitting threads. Another application is in the lead screws and feed screws of precision machine tools. For example, the lead screw, cross-feed screw, and compound rest feed screw of most lathes have Acme threads. Lead screws for certain types of presses and vises also use this type of thread.

Lathe threading tools are ground to the form of the thread desired. Figure 27.2 shows the proper tool forms for cutting external and internal Acme threads. The forms must be checked with an Acme thread gage (Fig. 27.3) during the grinding process of the bit, and also during the threading operation.

The various steps in cutting an Acme thread are similar to other types of threads; that is, set the compound rest at 14½ degrees and advance the compound feed after each cut, returning the cross feed each time to the same setting. Threading for Acme screw threads requires a lighter cut than for conventional 60 degree v threads because the total cutting face of the tool is longer.

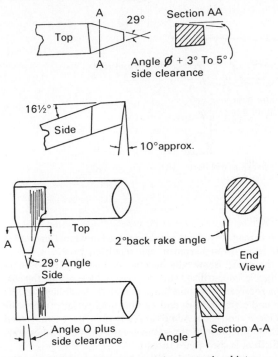

FIG. 27.2 Proper tool forms for cutting external and internal Acme threads. (*South Bend Lathe Co.*)

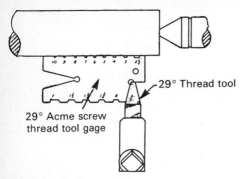

FIG. 27.3 Checking tool form with Acme thread gage. (*South Bend Lathe Co.*)

Square Threads

The square thread (Fig. 27.4) is seldom used any more because it is difficult to cut. Furthermore, the resulting thread is not as strong as the Acme. Some rifle barrels, such as the 1917 Enfield and 1903 Springfield, utilized square threads for screwing the barrel into the receiver. It is sometimes currently used, however, for many vise and clamp screws and similar applications. When the machinist has a choice, the Acme thread is the type recommended for all such applications; it is stronger, easier to cut, and capable of closer fit than is the square thread.

In cutting a square thread with a large lead, the tool angles must be absolutely correct. Clearance should be allowed on two sides, tapering from both the top and front of the tool as shown in Fig. 27.5.

External square threads should be cut to the minor diameter

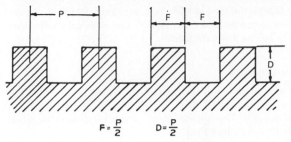

$$F = \frac{P}{2} \qquad D = \frac{P}{2}$$

FIG. 27.4 Square thread. (*South Bend Lathe Co.*)

plus about .005 in. The additional .005 in allows a small clearance at the bottom of the thread, which helps to compensate for any small inaccuracies in the tool or the cutting operation.

The tool must be fed directly into the work with the cross feed (or compound rest feed), and care must be exercised to avoid chatter and "hogging in." The best method is to set the compound

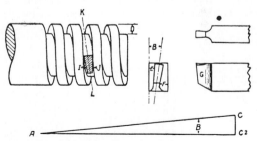

FIG. 27.5 Clearance for tool. (*South Bend Lathe Co.*)

rest at 0 degree and feed in with the compound; then back out and return the tool with the cross feed. Take very light cuts when turning or boring a square thread.

Whitworth Threads

A cross section of a typical Whitworth thread is shown in Fig. 27.6. This is a standard form in Britain and Europe for many types of threads. The smaller sizes of the Whitworth form are called British Standard Fine.

A Whitworth thread is cut in much the same manner as an Acme thread, except there are two major differences: the thread angle is smaller, and the radius at the top and bottom of the thread must be shaped properly with a formed tool.

Metric Threads

The Metric Standard screw thread shown in Fig. 27.7 is now the universally accepted thread form. The metric thread angle and form are identical to that of the National Form thread, and the cutting operation is exactly the same, with one exception: the lathe motor must be reversed after each cut. This procedure is

P = pitch = $\dfrac{1}{\text{no. threads/in}}$

D = depth = P $\times$ 0.6403

R = radius = 0.1373 P

FIG. 27.6 Typical Whitworth thread.

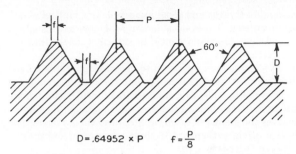

$$D = .64952 \times P \qquad f = \frac{P}{8}$$

FIG. 27.7 Metric standard screw thread. (*South Bend Lathe Co.*)

necessary because metric threads have no definite relation to the threading dial.

The following threading method applies to metric threads and to special fractional threads, wire feeds, and others. After the half-nut lever on the carriage is engaged for the first cut, it should not be moved until the thread has been completed. As the tool reaches the end of each cut, back out the cross feed, stop the lathe, and reverse the motor until the tool has been returned to the starting position. Then advance the cross feed to its original 0 position, turn in the compound rest feed for the next cut, start the motor, and repeat the cutting operation.

Multiple Threads

When fast travel is desired with a fine thread, a multiple thread may be used to reduce the original lead distance into two, three or more threads as required. This type of thread can be cut by two methods. The threading dial is quick, simple, and accurate for some double threads and some quadruple or "multiple-four" threads. In cutting, however, a larger tool clearance angle is required and the tool clearance must be sufficient for the lead, not merely the pitch.

Figure 27.8 shows the terms used to describe any thread. Lead is the distance the thread advances in one revolution; pitch is the distance between threads. In a simple thread, lead and pitch are the same. In multiple threads, the lead multiplied by the number of starts gives the pitch. For example, an 8-lead thread having two grooves is 16-pitch, or there are 16 threads per inch, and the distance between threads is ⅟₁₆ in.

When multiple threads are cut in the lathe, the first thread is cut to the desired depth. The work is then revolved part of a turn, and the second thread is cut, etc. In order to obtain spacing, it is advisable to mill as many equally spaced slots in the faceplate for the lathe dog as there are multiple threads to be cut—for a double thread, two slots; a triple thread, three slots, etc. If it is not convenient to cut slots in the faceplate, equally spaced studs may be attached to the faceplate and a straight-tail lathe dog used.

Another method for indexing the work when cutting multiple

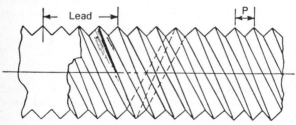

FIG. 27.8 Thread terms. (*South Bend Lathe Co.*)

threads is to disengage the change gears after the first thread has been completed and turn the spindle to the required position for starting the next cut.

When using the lathe dog for spacing, assume that a double thread is desired with a lead of 8 and a pitch of 16. The threading gears are set up for the lead. In setting up, always gear for the lead and not the pitch. Then proceed as for any single thread, but cut only to the depth of a 16-pitch thread. A gear table shows this to be .047 in or, if a regular 29 degree angle feed on the compound is used, the in-feed will be .054 in. After cutting the first thread to full depth, back up the head stock and fit the tail of the lathe dog in a slot opposite to the one used for the first thread. Back the compound to the original starting position and cut the second thread just the same as the first. If the 29 degree compound feed is used, the second thread will start alongside the first thread, and successive cuts will complete the thread. With the four slots of the faceplate, a quad thread may be cut. If a triple thread is desired, tap three equally spaced holes in the faceplate for studs and use a straight-tail dog as discussed previously.

Using the Thread Dial

Most lathes are fitted with a thread dial which may be used to cut multiple threads. In addition to the engagement points marked, the lathe can also be engaged at points indicated by dotted lines.

On any thread dial, numbered divisions indicate 1 in of carriage travel. One-half a numbered division is ½ in of carriage travel. The dial may be further subdivided to ⅛ in. With this information, consider a 6-lead thread. In 1 in of carriage travel, the cutting tool will engage at the first and the sixth threads. At ½ in of carriage travel, the tool will hit the third thread, but if the dial is used for ¼ in of carriage travel, the tool will hit at 1½ threads and will split the thread. Engagement at ⅛ in of carriage travel will again split it, producing a quad thread.

Setting Threads with the Compound Rest

This method is sometimes employed in the machine shop, especially for roughing the thread for 3, 5, and 6 starts which may be cleaned up later by chasing with a tap of proper pitch. The compound rest is set parallel. For a 6-start and 4-lead thread, with 24 pitch, the lathe gears are set for a 4-lead cut, and the thread is cut to a depth of a 24-pitch thread with successive plunge cuts made by in-feeding the cross slide. Then the compound is moved back a distance equal to the pitch, in inches, of a 24-pitch thread. A table shows this to be .042 in. The remaining threads are cut in the same manner. Threads cut by this method may show a slight variation in pitch, but this is equalized readily by running a few chasing cuts with a 24-pitch tap.

CUTTING THREADS ON A LATHE

To demonstrate the procedure of conventional thread cutting, assume that a 14-pitch fine thread is to be cut on the end of a 1 in diameter shaft—that is, the outside thread diameter is 1 in, and 14 threads are cut to every inch of the work. National Fine threads have a 60 degree angle. From a table or calculations, the depth of this thread, in this case, will be .0464 in.

To begin, the work is set up between centers in the lathe, after center drilling to obtain the correct-size center holes in the workpiece. Alternatively, a three-jaw chuck may be used for holding short lengths, but the thread will not be as "true" as it would if turned between centers.

Although the exact technique may vary between machinists, the next logical step is to undercut the shaft at the shoulder of the threads so that the cutting tool will not hit the edge or shoulder of the nonthreaded portion of the shaft. This undercut must be at least as deep as the finished thread to give full clearance.

The threading tool may be set to feed straight into the work, but better threads usually develop if the compound is set over 29 or 30 degrees. The tool then moves along the hypotenuse of a triangle, or 1.154 times the depth. See Fig. 27.9.

The threading tool should be checked with a thread gage to ensure that the tool is properly ground for cutting 14 threads per inch. The tool is then secured in the tool holder so that its point is at the exact center of the work. If placed either above or below this centerline, it will probably cut threads having a poor finish. Furthermore, a center gage is used to be sure the tool is square with the surface to be threaded. With the gage flat against the

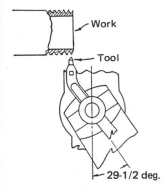

FIG. 27.9 Threading tool positioned at 29.5 degrees. (*South Bend Lathe Co.*)

work, the tool holder is moved until the tool fits in the 60 degree notch.

With the compound rest set at zero to provide a point from which to measure tool travel, the cross slide is advanced until the tool touches the work, and a stop is clamped to it at this setting.

The gearing between the lead screw and lathe spindle is set for 14 threads per inch. The lathe backgear is also engaged to slow the spindle down to a very slow speed, say 20 rpm.

The carriage is engaged to the lead screw by the carriage engagement lever and the threading dial is used to locate the proper place to reengage the carriage for subsequent cuts. To test the setup, a light trial cut is made with the tool advanced about .002 in by the compound rest feed. This will cut light lines in the work, giving an easy way to count the threads per inch.

As soon as the tool enters the undercut, the carriage is disengaged from the lead screw, and the tool is backed out by means of the cross feed. The tool goes back to the starting position. After the cross slide has been moved against its stop, the tool is advanced about .005 in by the compound rest feed. When the total movement of the compound rest is .054 in, the thread is finished.

To obtain a smooth thread, a good cutting oil is used. It is also important to engage the lead screw some distance before the tool enters the cut so as to compensate for any backlash in the gears. As the thread nears its specified depth, it can be checked with a thread micrometer. To be sure of a fit, it is often advisable to test the thread directly on the part for which it is made.

28

Hand Files and Filing

Files and rasps have three distinguishing features: length, kind (or name), and their cut. A file's length is always measured exclusive of the tang; the kind refers to the file's shape or style, and the cut refers to both the character and the relative degrees of coarseness of the teeth. See Fig. 28.1 for a hand file's basic parts.

FILE FEATURES

Length

The length of the file is the distance between its heel and tip. The tang is never included in the length. In general, the length of a file bears no fixed proportion to either the width or the thickness, even though the files are of the same kind.

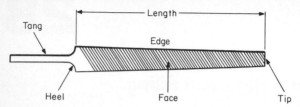

FIG. 28.1 Basic parts of a hand file.

Kind

By the kind we mean the various shapes or styles of files, as distinguished by such technical names as flat, mill, half-round, etc. These are divided, from the form of their cross sections, into three general geometrical classes: quadrangular, circular, and triangular. From them are derived, further, odd and irregular forms, and cross sections which are classified as miscellaneous. As a file increases in length it grows in cross-section size. These sections, in turn, are subdivided, into taper and blunt.

Taper designates a file, the point of which is more or less reduced in size (either in width or thickness, or both) by a gradually narrowing section extending from one-half to one-third the length of the file, from the point.

Blunt designates a file that preserves its sectional size throughout, from point to tang.

Cut

The cut of files is divided, with reference to the character of the teeth, into single, double, rasp, and curved, and with reference to the coarseness of the teeth, into coarse, bastard, second, and smooth cuts. See Fig. 28.2.

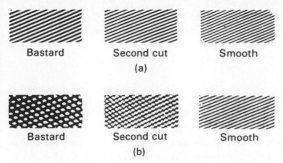

FIG. 28.2 Files have either single- or double-cut file teeth. They also come in various sizes, the most common being, bastard, second cut, and smooth. (*a*) Single cut and (*b*) double cut.

The single-cut type of file has a single series of teeth. This type of file is usually used with a light pressure to produce a smooth surface finish, or a keen edge on a knife, shears, sawtooth, or other cutting implement.

Double-cut files have two series of diagonal teeth. The first of its cuts is called the overcut, the second the upcut. They are usually used under heavier pressure, for fast metal removal and where a rougher finish is permissible.

The rasp cut is a series of individual teeth produced by a sharp, narrow, punch-line cutting chisel. It is an extremely rough cut and is used principally on wood, leather, hoove, aluminum, lead, and similarly soft substances for fast removal of material.

For unusual types of files, such as those used on the flat surface of aluminum and steel sheets, a special curved tooth (single) is used.

Regarding coarseness, the coarse and bastard cuts are used on the heavier classes of work; the second and smooth cuts for finishing or more exacting work.

TYPES OF FILES

Mill Files

Mill files are so named because they are widely used for sharpening mill or circular saws. These files are also useful for sharpening large crosscut saws and mowing-machine knives; for lathe work and draw filing; for work on compositions of brass and bronze; and for smooth-finish filing in general.

Mill files are single-cut and are tapered slightly in width for about a third of their length; 12-, 14-, and 16-in files are also tapered in thickness. They are usually made with two square edges, with cuts thereon as well as on the sides. They can also be made with one or two round edges to maintain rounded gullets on crosscut saws. The mill blunt is likewise used for crosscut saws and bucksaws, as well as for general filing.

Machinist's Files

Machinist's files, as the name indicates, are widely used by machinists in repair and manufacturing shops, and wherever metal must be removed rapidly and finish is of secondary importance. They include flat, hand, round, half-round, square, pillar, three-square, warding, knife, and a number of less commonly known kinds. With certain exceptions in round and half-round, all machinist's files are double-cut.

Swiss Pattern Files

The so-called "Swiss pattern" files constitute a vast field of their own. They are used by tool and die makers, jewelers, model makers, delicate instrument parts finishers, and home people. In short, everyone who does superfine precision filing will have many uses for Swiss pattern files.

Swiss pattern files are made to more exacting measurements than the conventional American pattern files. Although some cross sections of both types are similar, the shapes differ. The points of Swiss pattern files are smaller, and the tapered files have longer tapers. They are also made in much finer cuts, which vary from no. 00, the coarsest, to no. 6, the finest.

Swiss pattern files are primarily finishing tools used for removing burrs left over from previous finishing operations; trueing up narrow grooves, notches, and keyways; rounding out slots and cleaning out corners; smoothing small parts; and doing the final finishing on all sorts of delicate and intricate pieces.

Swiss pattern files come in upwards of a hundred shapes—many of them in a range of sizes. In addition to the more common named ones such as Swiss pattern half-round, Swiss pattern square, Swiss pattern round, Swiss pattern pillar-narrow, Swiss pattern three-square, there are other classifications such as square- and round-handle needle, die sinker, parallel and bench-filing machine, broach corrugating, joint, pippin, slitting, crossing, screw hand, equaling, warding, barrette, and silversmiths' riffler.

Curved-Tooth Files

Curved-tooth files cover a distinct filing field and have a considerable range of shape and structural characteristics. They are widely used in the automobile manufacturing and repairing indistries for work on aluminum and sheet steel (on flat or curved surfaces). They are also used on such soft metals as brass and babbitt, and often on iron and steel. Because of their curved teeth, they readily clear themselves of chips and have the correct rake for speed and economy.

Regular curved-tooth files are made in both rigid and flexible forms. The rigid form is either tanged for the conventional handle or plain (with hole at each end) for special holders. The flexible form is plain only. Curved-tooth files are available for standard, fine, and smooth cuts, and in parallel-flat, square, pillar, pillar-narrow, half-round, shell, and molding types.

FILE SELECTION

There are thousands of kinds, cuts, and sizes of files. That is because there are thousands of different filing jobs, each of which can be done better by using the right file for the job.

The right file allows doing the job properly, whereas the wrong one does not—and often, in fact, ruins the work. The right file saves time, because it performs correctly, and usually faster, on the kind of metal or work for which it is designed. The right file permits a greater number of efficient filing strokes—per file and per file cost. Sum up all these advantages and they represent a big item of savings in a day's filing and production costs.

Many factors enter into the selection of the right file for the job. In general, it may be said that different files are required (1) to file a flat or convex surface; (2) to file a curved or concave surface; (3) to file and edge; (4) to file a notch, a slot, or a square or round hole.

But these factors can immediately become complicated by: (1) the kind of metal or other material to be filed; (2) the kind, shape, and hardness of object or part of be filed; (3) the location, size, and character of the surface, edge, notch, slot, or hole to be filed; (4) the amound of metal to be removed—and the practical time permitted for removing it; and (5) the degree of smoothness or accuracy required.

All these conditions have a bearing on the kind, size and cut of file which will best attain a particular objective. Calculate the number of possible combinations of such conditions, and selecting exactly the right file for any combination would seem to be a sizable task for one person.

USING FILES

Skill or aptitude of the person is usually shown by the way a person masters the fundamentals of a trade. Hammer, saw, chisel, plane, file—there is a right and a wrong way to use each. Furthermore, each calls for different "touches" according to the character of the work, the working conditions and the kind of results sought. The logic applies with particular aptness to files and how to get the most out of them.

There are three elemental ways in which a file can be put to work:

1. *Straight filing.* This consists of pushing the file lengthwise—straight ahead or slightly diagonally—across the work (since all files, with the exception of a few machine-operated files, are designed primarily to cut on the forward stroke).

2. *Draw filing.* This consists of grasping the file at each end and pushing and drawing it across the work.

3. *Lathe filing.* This consists of stroking the file against work revolved in a lathe.

Most work to be filed is held in a vise. For general filing, the vise should be about elbow height. If a great deal of heavy filing is to be done, it is well to have the work lower. If the work is of a fine or delicate nature, it should be raised near the eye level.

For work which is apt to become damaged by pressure when held in a vise, it is well to provide a pair of "protectors"—pieces of zinc, copper, or other fairly soft metal placed between the jaws and the work to be held. For holding varying sizes of round pieces such as small rods and pins, a block of hard, close-grained wood with a series of different-sized groves is sometimes used where many such pieces must be filed.

With files intended for operation with both hands, one of the most generally accepted ways of grasping the handle is to allow its end to fit against the fleshy part of the palm below the joint of the little finger, with the thumb lying parallel along the top of the handle and the fingers pointing upward toward the operator's face.

The point of the file is usually grasped by the thumb and the first two fingers of the other hand. The hand may then be positioned to bring the thumb, as its ball presses upon the top of the file, in line with the handle when heavy strokes are required.

When a light stroke is wanted, and the pressure demanded becomes less, the thumb and fingers of the point-holding hand may change their direction until the thumb lies at a right angle, or nearly so, with the length of the file, the positions changing as needed to increase the downward pressure.

In holding the file with one hand, as in filing pins, dies, and edged tools not held in a vise, the forefinger—instead of the thumb—is generally placed on top and as nearly as possible in the direction of the file's length.

The most natural movement of the hands and arms is to "carry" (stroke) the file across the work in curved lines. This tends toward a rocking motion and, consequently, a convex surface where a level surface is desired.

For the usual flat filing, the operator should aim to carry the file forward on an almost straight line, changing its course enough to prevent "grooving."

One of the quickest ways to ruin a good file is to use too much pressure—or too little—on the forward stroke. Different materials, of course, require different touches; but, in general, just enough pressure should be applied to keep the file cutting at all times. If allowed to slide over the harder metals, the teeth of the file rapidly become dull; and if they are overloaded by too much pressure, they are likely to chip or clog.

On the reverse stroke it is best to lift the file clear off the work, except on very soft metals. Even then the pressure should be very light—never more than the weight of the file itself.

Draw Filing

Draw filing consists of grasping the file firmly at each end and alternately pushing and pulling the file sidewise across the work. Since files are primarily made to cut on a longitudinal forward stroke, a file with a short-angle cut should never be used for draw filing, because of the likelihood of scoring or scratching instead of shaving or shearing. When properly done, draw filing produces a somewhat finer finish than "straight" filing.

Draw filing is used extensively in file factories themselves, in preparing file blanks for cutting. It assures a perfectly smooth, level surface and uniform file teeth.

Ordinarily, a standard mill bastard file is used for draw filing. But where a considerable amount of stock is to be removed—as on the edge of a metal sheet or plate—a flat or hand file (double-cut) will work faster. However, the double-cut file usually leaves small ridges in the work and consequently does not produce a finished job where a smooth surface is required. In such cases the double cut may be used for the roughing down, then be followed by the single-cut (mill) for finishing. One kind of application for draw filing is shown in Figs. 28.3 and 28.4.

Lathe Filing

When file is held against work revolving in the lathe, it should not be held rigidly in a stationary position but should be stroked constantly. A slight gliding or lateral motion helps the file to clear itself of chips and avoids producing ridges or scores.

While the ordinary mill file is normally capable of doing good lathe-filing work, there is a special long-angle lathe file with teeth cut at a much longer angle than those of the standard mill file. This provides a much cleaner shearing, self-clearing file, eliminates drag or tear, overcomes chatter, and reduces clogging. It is very fast cutting, and the most delicate touch brings the work to a fine silken-smooth finish. "Safe" (uncut) edges on this file protect any shoulders of the work which are not to be filed. They also protect the dog which enables the lathe to revolve the work.

Lathe filing is most commonly employed for the purpose of fitting shafts. In general, the highest spindle speed should be used on such work; and where the amount of stock to be removed is considerable, the 12- or 14-in long-angle lathe file is the preferable size. This type file is largely used in industrial plants to bring a shaft down to a drive fit. For a running fit, for example, where a shaft is to run on a bearing, a mill file will best provide the necessary smooth finish. Where a fine, highly polished finish is

FIG. 28.3 Draw filling a Stevens single-shot rifle receiver to remove rust pits prior to polishing for reblueing.

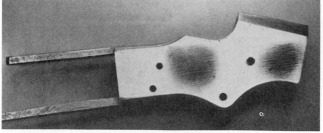

FIG. 28.4 Receiver after draw filing. Note the "low" area on the side of this receiver, it shows that the surface is not a perfect plane.

desired, a Swiss pattern hand or pillar file no. 4 or no. 6 cut may be used with very good results.

Many lathe filers make a practice of running a new file over a flat piece of cast iron before using it on lathe work, for the purpose of removing extreme sharpness from the top edges of the teeth. This is necessary, however, only on work requiring a very smooth finish.

In using the long-angle lathe file, care should be taken at shaft ends and shoulders, as this fast-cutting file may easily cut too deeply at such points. It is also important never to run the hand over work on a lathe, as the accumulated oil and moisture from the hand will sometimes so coat the surface of the work that it becomes difficult for the file to take hold again.

In lathe-filing work which does not have cylindrical surface, but an oval, elliptical, or irregularly rounded form, the finer or lighter cutting Swiss pattern files will be found most satisfactory. They are made in a wide range of shapes, sizes, and cuts, and will impart a smooth finish to any work filed in a lathe.

THE CARE OF FILES

File life is greatly shortened by improper care as well as by improper use and improper selection. Files should never be thrown into a drawer or toolbox containing other tools or objects. They should never be laid on top of or stacked against each other. Such treatment ruins the cutting edges of their teeth. Keep them separate, standing with their tangs in a row of holes or hung on a rack by their handles. Keep them in a dry place so rust will not corrode their tooth points.

It is also of great importance to keep files clean of filings, or chips, which often collect between the teeth during use. After every few strokes the good mechanic taps the end of the file on the bench to loosen these chips. The mechanic always has on hand a file card or brush. The teeth of the file should be brushed frequently with this type of cleaner, and always before the file is put away. To remove obstinate "pinnings" which sometimes clog up the teeth and cause scratches on the work, a "scorer"—made of soft iron—is often a further help. Oil or grease on the file should be removed with chalk. A file kept clean lasts longer and does better work.

File cleaners are made in two styles: file card for more general uses and file brush, combining brush and card, for the finer-cut files.

29

Mechanics

Energy, in brief, means ability to do work. Mechanical work is done when any kind of energy is used to produce motion in a body formerly stationary, or to increase the rate of motion of a body, or to slow its rate of motion.

The most generally used unit of work is the foot-pound, which represents a mass of 1 lb lifted 1 ft against the force of gravity. The total amount of work done is equal to the number of feet of motion multiplied by the number of pounds moved. Time does not enter into the matter of mechanical work.

MECHANICAL POWER

Power is the rate of doing work involving work and time. One of the units in which power may be measured is foot-pounds per second. Power is equal to the total amount of work divided by the time taken to do the work. It is assumed that the work is being done at a uniform or constant rate, at least during the period of time measured.

The foot-pound per second is a unit too small to be used in practice. Mechanical power most often is measured in the unit called "horsepower." One horsepower is the power rate corresponding to 550 ft-lb/s. Since there are 60 seconds in a minute, 1 hp also corresponds to 550 times 60, or to 33,000 ft-lb/min.

ELECTRIC POWER

For an electric motor to continue working at the rate of 1 hp, an electric current must be sent through the motor at a certain number of amperes when the potential difference across the motor terminals amounts to a certain number of volts. The number of amperes and the number of volts would have to be such that multiplied together they would equal 746. A unit of electric power to describe this product of amperes and volts is called the watt. One watt is the power produced by a current of 1 ampere when the pressure difference is 1 volt.

The total number of watts of power is equal to the number of amperes of current multiplied by the number of volts of potential difference, both with reference to the device in which power is being produced. The number of watts must be 746 to produce 1 hp, so 746 W of electric power is equivalent to 1 mechanical horsepower.

One watt-hour of electric energy is the quantity of energy used with a power rate of 1 W when this rate continues for 1 hour. That is, the watt-hours of energy are equal to the number of watts multiplied by the number of hours during which power is used at this rate. A 60-W electric lamp uses power at the rate of 60 W so long as it is lighted to normal brilliancy. But the total quantity of energy used by the lamp depends also on the total length of time it remains lighted. If the 60-W lamp is kept lighted for 10 hours, it will have used 60 × 10, or 600 watt-hours of electric energy.

ELECTRIC MOTORS

Electric motors are machines that change electrical energy into mechanical energy. They are rated in horsepower. The attraction and repulsion of the magnetic poles produced by sending current through the armature and field windings cause the armature to rotate. The armature rotation produces a twisting power called torque. Electric motors are a common sight in the machine shop, because practically all power tools are operated by electric motors. Therefore, every machinist should have at least an elementary knowledge of electric motors and their application. The following information should prove invaluable to the machinist for use in selecting various electric motors to power machine shop tools.

Style number.   Identifies that particular motor from any other.

Serial data code.   The first letter is a manufacturing code used at the factory. The second letter identifies the month, and the last two numbers identify the year of manufacture. For example, D85 is April 1985.

Frame.   Specifies the shaft height and motor mounting dimensions and provides recommendations for standard shaft diameters and usable shaft extension lengths.

Service factor.   A service factor is a multiplier which, when applied to the rated horsepower, indicates a permissible horsepower loading which may be carried continuously when the voltage and frequency are maintained at the value specified on the nameplate, although the motor will operate at an increased temperature.

Phase.   Indicates whether the motor has been designed for single- or three-phase service. It is determined by the electrical power source.

Degrees C ambient.   The air temperature immediately surrounding the motor. Forty degrees Celsius is the NEMA maximum ambient temperature.

Insulation class.   The insulation system is chosen to ensure the motor will perform at the rated horsepower and service factor load.

Horsepower.   Defines the rated output capacity of the motor. It is based on breakdown torque, which is the maximum torque a motor will develop without an abrupt drop in speed.

rpm.   Speed in revolutions per minute. The rpm reading on motors is the approximate full-load speed.

Amps.   Gives the amperes of current the motor draws at full load. When two values are shown on the nameplate, the motor usually has a dual voltage rating. Volts and amps are inversely proportional; the higher the voltage, the lower the amperes, and vice versa. The higher amp value corresponds to the lower voltage rating on the nameplate. Two-speed motors will also show two ampere readings.

Hertz.   Just about everything in the United States is serviced by a 60-Hz alternating current. Therefore, most applications will be for 60-Hz operations.

Volts.   Volts is the electrical potential "pressure" for which the motor is designed. Sometimes two voltages are listed on the nameplate, such as 120/240. In this case the motor is intended for use on either a 120- or 240-volt circuit. Special instructions are usually furnished for connecting the motor for each of the different voltages.

kVA Code. This code letter is defined by NEMA standards to designate the locked-rotor kVA per horsepower of a motor. It relates to starting current and selection of fuse or circuit breaker size.

Housing. Designates the type of motor enclosure. The most common types are open and enclosed. Open drip-proof has ventilating openings so constructed that successful operation is not interfered with when drops of liquid or solid particles strike or enter the enclosure at any angle from 1 to 15 degrees downward from the vertical. Open guarded has all openings giving direct access to live metal or hazardous rotating parts so sized or shielded as to prevent accidental contact as defined by probes illustrated in the NEMA standard. Totally enclosed motors are constructed to prevent the free exchange of air between the inside and outside of the motor casing. Totally enclosed fan-cooled motors are equipped for external cooling by means of a fan that is integral with the motor. Air-over motors must be mounted in the airstream to obtain their nameplate rating without overheating. An air-over motor may be either open or enclosed.

Hours. Designates the duty cycle of a motor. Most fractional horsepower motors are marked continuous for around-the-clock operation at the nameplate rating in the rated ambient. Motors marked "one half" are for ½-hour ratings, and those marked "one" are for 1-hour ratings.

The following terms are not normally found on the nameplate, but are important considerations for proper motor selection.

Sleeve bearings.   Sleeve bearings are generally recommended for axial thrust loads of 210 lb or less and are designed to operate in any mounting position as long as the belt pull is not against the bearing window. On light-duty applications, sleeve bearings can be expected to perform a minimum of 25,000 hours without relubrication.

Ball bearings.   These are recommended where axial thrust exceeds 20 lb. They too can be mounted in any position. Standard and general-purpose ball bearing motors are factory-lubricated and under normal conditions will require no additional lubrication for many years.

Rigid mounting.   A rectangular steel mounting plate which is welded to the motor frame or cast integral with the frame—the most common type of mounting.

Resilient mounting.   A mounting base which is isolated from motor vibration by means of rubber rings secured to the end bells.

Flange mounting.   A special end bell with a machined flange which has two or more holes through which bolts are secured. Flange mountings are commonly used on such applications as jet pumps and oil burners.

Rotation.   For single-phase motors, the standard rotation, unless otherwise noted, is counterclockwise facing the lead or opposite shaft end. All motors can be reconnected at the terminal board for opposite rotation unless otherwise indicated.

MACHINE DRIVES

Most machines in use today are driven by electric motors. For an electric motor to perform efficiently, the proper type of motor and control for the application must be selected. To make such a selection, a knowledge of the characteristics of ac and dc motors and controls is essential. It is also necessary to have a knowledge of the general operating characteristics of the load the motor will "pull" or operate, such as power and speed requirements, special operating conditions, control features, and similar items.

Types of Drives

In general, two methods of drive have been traditional: using one motor to drive two or more machines as a group or using an individual motor for each machine. While the former was the popular method decades ago, the latter method is the most popular today. In the group drive system, the various machines are connected together by shafting and belts, although chains, gears, or other mechanical devices may be used (Fig. 29.1).

Seldom will a group drive system be seen in any modern machine shop. Practically all systems are now of the individual drive type. The reasons are many, but flexibility of location or arrangement of the machine with individual motor drives makes it possible to place each machine in the best position to suit the flow of material and save handling and trucking. Also, as shop or work conditions change, each machine—complete with drive and control as a unit—may be readily moved to some other location. The elimination of overhead belting and shafting greatly improves the lighting on the machines and also facilitates the use of material-handling equipment such as small cranes and similar items. The most important factor, however, is probably the ease with which machines using individual motors may be started and stopped. Also, with this system, the exact speed required may be obtained and maintained indefinitely, being independent of the load on other machines and of temperature and atmospheric conditions, which may affect belt transmission.

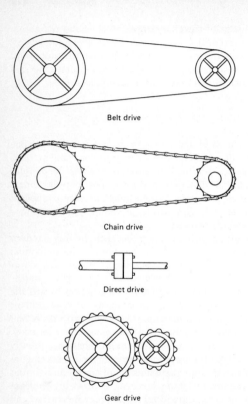

Belt drive

Chain drive

Direct drive

Gear drive

FIG. 29.1 Types of drives for machine tools.

SELECTING THE MOTOR FOR THE JOB

When selecting motors for a particular application, no set rules will fit each and every application, but there are some important machine and load characteristics that should be known for practically all requirements:

1. The speed requirement of the driven shaft, in rpm
2. The range of speed required, if the machine is adjustable
3. The horsepower required at maximum speed or loading
4. If the load is not constant, the cycle duty, including variable items, such as load, time, speed, weights, and other factors
5. If the speed varies, the variation of the torque with the speed
6. Torque: starting, pull-in, pull-out, or maximum—all in percent of full-load values
7. Mechanical connection: belt, chain, coupling, or gear.

Determining Torque

It is sometimes desirable to check the manufacturer's data on the motor characteristics. This is done by installing a temporary motor and taking power readings under various operating conditions.

The required starting torque can be determined by wrapping a rope around the driven pulley and then measuring, with a spring scale, the pull which will start the machine and turn it over. The starting torque, in pound-feet, is the product of the reading on the scale, in pounds, and the radius of the pulley, in feet. For example, assume that it requires a pull of 75 lb on a driven pulley of 6-in radius to start the motor. The starting torque required is

$$75 \times 6 \;/\; 12 = 3.75 \text{ lb-ft at 1-ft radius}$$

In most cases, however, the manufacturer's data can be relied upon and will suffice for most applications when a motor is selected for a particular job.

Power Supply

If the power supply is alternating current, it is necessary to know the frequency, voltage, and number of phases. However, if the power supply is direct current, only the voltage need be known.

The characteristics of the electric service and its limitations must be considered in every instance when motors are selected for a given application.

Load Factor

The load factor is the ratio of the average load to the maximum load over a certain period. The time may be either the normal number of operating hours per day or 24 hours. The average load is equal to the kilowatt-hours used in the specified time, as measured by watt-hour meters, divided by the number of hours. The maximum load is the highest load at any one time as measured by some form of maximum-demand or curve-drawing watt-hour meter.

For comparison of industrial loads, the maximum load is taken as the load which would be obtained if all motors were operating continuously at the full-rated load for the same period. This input to the motors is obtained from the equation:

Full-load input, in kW = total hp rating of motors
$\times$ .746 / average efficiency of motors $\times$ hours of operation per day

If the load factor is based on the number of working hours per day, the 24-hour load factor may be obtained by multiplying the given load factor by the number of hours and dividing by 24.

Types of Mounting and Enclosures

Two important mechanical characteristics must be considered when selecting motors for a particular machine-shop application. One is the type of mounting to be used, and the other is the type of enclosure and ventilation.

The types of motor mounting to consider are the horizontal, vertical flange, gear motor, and others. Common types of enclosures include open drip-proof, totally enclosed fan-cooled, totally enclosed explosion-proof, and separately vented—pipe-vent or blower.

Open drip-proof.   This design draws outside cooling air into the motor for ventilation. It is primarily for clean, dry areas indoors. Contaminated cooling air will normally reduce the life of the motor winding and bearing grease. This motor is good for general-purpose use but not ideal where little maintenance can be performed.

Totally enclosed fan-cooled.   Outside cooling air is directed over the motor by an exterior shaft-mounted fan. It is ideal for dusty atmospheres and many hostile areas. It is primarily used in areas where the motor may not be accessible for maintenance. Cooling is not as efficient as in an open-type motor.

Totally enclosed explosion-proof.   Ventilation here is the same as in the totally enclosed fan-cooled. The motors are specifically designed for installation in hazardous areas as defined by the NE Code. The motor is built to contain and withstand an explosion within its own enclosure. Because highly flammable gases and dust require special designs, attention is given to machined fits and mating surfaces to ensure adequate flame paths to contain and extinguish flames or sparks before they reach the outside air.

Separately vented-pipe or blower.   These motor types are the least common. Blower cooling is used where high-heat-generating duty cycles are found and/or motor size is at a premium. Blower-forced air cooling can remove heat quickly. This design approach can be expensive compared to conventional enclosures and is dependent on heat removal as the key design factor rather than torque limitations. Pipe ventilation is similar except here the motor is generally ducted to a supply of clean, fresh air. It may or may not have a blower. Often the motor may require oversizing where the duct is long. Maintenance is critical to each of these to ensure that ventilation passages are clear. The application considerations that follow can be applied to the basic enclosures outlined.

High-Speed Belting

An ever-increasing number of processes require high speed for greater output or simply to properly blend special materials being mixed. This, coupled with the fact that higher-capability fiberglass belts are often used, can reduce bearing and shaft life.

High-speed belting usually employs a 3600-rpm motor. Conventional bearings operating at high speed and high load factors begin to approach their design point in limiting speed. Lubrication breakdown is also prevalent.

Drives as a minimum should utilize ductile iron, dynamically balanced sheaves, and matched belts. Transmittal of this information to the motor supplier with the belt drive details is critical if a marriage of the motor drive is to be successful.

The bearing should be selected to maximize life, and this can mean a standard ABEC-1 commercial-motor-quality standard bearing, an ABEC-3 for truer tolerances on the ball and shaft bore, or a bronze or phenolic retainer for the balls to increase speed capability. It is critical not to neglect the limiting bearing speed recommended by the bearing manufacturer.

Lubrication is generally grease of a no. 2 grade. Oil or oil-mist lubrication may be considered, although grease is a more economical approach and is the first choice.

Shaft deflection imposed by heavy belt tensions or overtensioning can mean that oversize shafting will be required, as well as increased diametrical clearance between the shaft and bearing housing to prevent rubbing. A common conception is that changing to higher-strength shaft material will provide a stiffer, lesser deflecting assembly. Actually, deflection is determined by the relationship

$$\text{Deflection} = WI^3/48EI$$

where W = load
  I = length of shaft
  E = modulus of elasticity (stress/strain)
  I = moment of inertia

Diameter changes have the greatest impact upon shaft deflection; the modulus of elasticity is the same or nearly the same for almost all common motor shaft materials and has little effect upon deflection.

Recommendations

1. Check bearing life in hours for radial shaft load.
2. Verify bearing speed capability.
3. Use balanced, ductile sheaves and matched belts.
4. Check standard shaft deflection, and change diameter if necessary.
5. Open shaft-bearing housing clearance.

6. Be sure belt drive is tensioned properly (glass belts have reduced deflection rates for tensioning).

Duty Cycle

Because of special product requirements, motors are subjected to drag cycles—intermittent operation with frequent starting, stopping, and often reversing. The associated heating created by high loads for short periods of time often demands high insulation classes and/or blower cooling.

An approximate determination of required motor horsepower can be made from cyclic load curve using the rms (root mean square) method which is an arithmetic integration of the square of the load curve.

HP is the average horsepower between T_0 (time 0) and T_{10}, etc. After arriving at the final rms horsepower value, it is necessary to take the peak horsepower value encountered during the cycle and convert it to pound-feet. Then this value can be plotted on the motor speed-torque curve to be sure the motor can produce this torque without stalling.

Recommendations

1. Calculate rms horsepower.
2. Select a motor with one higher insulation class to cover peak loads.

There are numerous cases where reversing is a part of the duty cycle. Reversing capacity is rated in idle reversals that can be performed with no load or connected inertia. Since the reversing capacity of a motor is inversely proportional to the total connected inertia, the following applies:

$$R_x/R_i = WK_{2r}/WK_{2t}$$

where R_x = reversal acceleration with connected inertia
R_i = reversal acceleration idle
WK_{2r} = rotor inertia, lb-ft
WK_{2t} = total connected inertia (load and rotor)

Reversal capacity is limited by the amount of heat generated by the reversal itself in a particular frame size. Once the data in the preceding equation are determined, reversals can be calculated. All that remains is to assign values of heat units to the cycle. A heat unit is a segment of the cycle to which a heating increment is assigned:

- Motor acceleration to full speed = 1 heat unit
- DC brake stop = 1 heat unit
- Plug stop = 3 heat units
- Plug reversal = 4 heat units

By selecting the applicable cycle portions listed, total heat units can be easily calculated. Dividing the connected inertia reversal rate R_x by the heat units per cycle yields the loaded cycles per minute.

Recommendations

1. Low-inertia-rotors in high reversal rate design.
2. High stator copper content for lower motor temperatures.
3. Pinned or keyed rotors for mechanical resistance to reversing.
4. Aluminum vent fan keyed to shaft for positive cooling and mechanical resistance to reversing.

Inertia

High-inertia drive motors are those capable of accelerating very large loads from rest to full speed. Typically, these loads are fans or centrifuges.

The primary consideration is that of heat dissipation. During starting, heat is generated inside the motor rotor and stator. The degree and effectivenss of heat dissipation is a direct function of the magnitude of inertia that can be accelerated.

During starting, most heat is generated in the motor rotor and stator due to high inrush current and high rotor slip. Since this heat is directly related to the inertia accelerated, the motor must then be able to absorb the heat until heat transfer takes place, allowing normal ventilation to take over. Temperature is then directly proportional to pounds of active material (copper and steel). If acceleration time is increased by means of reduced voltage starting, then heat transfer can allow the motor to adequately dissipate the heat generated. Acceleration times will be lengthened considerably—often from 1 to 2 minutes up to 8 to 10 minutes.

The important criterion to consider here is that the available torque at reduced voltage must exceed the friction and windage torque of the drive by an ample amount to provide adequate acceleration. Torque becomes the important design factor rather than horsepower or start-stop requirements. Star-delta starting as compared to full-voltage starting would yield the different motor capabilities shown in Table 29.1.

Reduced-voltage starting will typically reduce stator temperature to as low as 50 to 60 percent of the line start value. This is often desirable for process requirements in air-conditioned facilities as well as for substantially extending motor life.

Thermal protection is highly desirable, and two sets of protectors are recommended. The first set of protectors is for starting.

TABLE 29.1 75-hp, 1800 rpm Enclosed Motor

Motor inertia	Star-delta start	Full-voltage start
Capability	6600	2600
Acceleration time	10 min	2 min

They are for the maximum temperature condition to eliminate nuisance tripping for two consecutive starts of the high inertia with the motor at operating temperature. The second set of protectors is for the running mode, and they are set for normal rated running temperatures. The motor insulation is rated for the maximum rated temperature condition so as to prolong life.

Recommendations

1. Define inertia, friction, and windage.
2. Use reduced voltage starting if inertia values are high (higher than across-the-line start).
3. Verify motor accelerating torque.
4. Define starting cycles.
5. Define duty cycle.

Glossary

Abrasive Grinding material such as sandstone, emery, carborundum, etc. The natural abrasives include the diamond, emery, corundum, sand, crushed garnet and quartz, tripoli, and pumice. The artificial abrasives are in general either silicon carbide or aluminum oxide, and are marketed under many trade names.

Absolute A term frequently used in the trades to indicate a thing as being perfect or exact.

Accurate Without error, precise, correct, conforming exactly to a standard.

Acetone An inflammable liquid with a bitter taste, obtained by the destructive distillation of certain wood, acetates, and various organic compounds.

Acetylene gas An illuminating gas resulting from the action of water on calcium carbide. Also used for oxyacetylene welding.

Acid bath Pickle used for cleansing metal objects in preparation for electroplating; usually used for dipping.

Acme thread A screw thread with a section between the square and threads. Used extensively for feed screws. The included angle of space is 29 degrees, as compared to 60 degrees of the National Coarse or U.S. thread.

Acute angle One which is less than 90 degrees, or less than a right angle.

Adhesion or adhesive power   The friction existing between a driving wheel and the surface with which it is in contact. The property which enables one surface to adhere or stick to another, as in glue practice.

Adjustable parallels   Wedge-shaped bars of iron placed with the thin end of one on the thick end of the other. The top face of the upper and the bottom face of the lower remain parallel, but the distance between the two faces can be increased or decreased, and the bars locked in position, by means of a screw.

Adjustable reamer   A reamer which can be increased in size, usually by means of a central bolt or screw. Tightening the bolt causes an expansion of the reamer.

Adjustable tap   A tap, usually made with inserted blades or chasers, capable of radial adjustment.

Adjusting screw   A setscrew, used to adjust the position of machine parts more accurately than would be possible by merely setting the parts to dimensions.

Adjustment   The placing and setting of engine or machine parts in related position.

Aerial metal   A very strong alloy of aluminum and lithium. It is very light, weighing only about 100 lb/ft^3.

Agitator   A mechanical stirring device, commonly used in mixing ingredients in large vats or tanks.

Air compressor   A machine in which air is compressed for use as motive power.

Air hardening   The hardening of high-speed steels by air blast.

Allen screws   Cap screws and setscrews having a hexagonal socket in the head. Such screws are adjusted by means of a hexagonal key.

Alloy   A homogeneous combination of two or more metals, usually a fine and baser metal. Examples are white metal and babbitt metal.

Alloy steel   A steel which is alloyed with one or more of the following metals: manganese, nickel, tungsten, molybdenum, vanadium, and chromium. These alloy steels are strong, tough, and hard.

Alumen A very strong aluminum alloy which can be forged and machined. It consists of 88 percent aluminum, 10 percent zinc, and 2 percent copper. It is heavier than aluminum.

American screw gage A standard gage for checking the diameter of wood screws and machine screws.

Angle iron A strip of structural iron, the section of which is in the form of a right angle.

Angle of repose or angle of friction The angle of a plane surface, inclined relatively to the horizon, upon which a body will, under specific conditions, just begin to slide. It varies with the nature of the particular materials placed in contact.

Angle plate Used in setting up work, generally for machinery. Formed of two plates of cast iron at right angles with each other, and pierced with holes or slots for the reception of bolts.

Angular cutter A milling cutter on which the cutting face is at an angle with regard to the axis of the cutter.

Anagular gears Bevel gears designed to run at angles other than a right angle.

Angular velocity The ratio which the arc described in one second, by a body or point rotating about a center, bears to the radius.

Annealing The gradual heating and the gradual cooling of glass, metals, or other materials to reduce brittleness and increase flexibility, etc.

Annular wheel A ring gear with teeth fixed to its internal circumference; also called an internal gear.

Antimonial lead An alloy of from 4 to 10 percent antimony with 90 to 96 percent lead. Used for storage-battery plates.

Antimony A silver-white, hard, crystalline metallic element related to arsenic and tin. Frequently used in alloys of tin and lead to give hardness.

Anvil A steel or iron block upon which forging is done.

Anvil block A massive block of cast iron which is placed beneath the anvils of steam and other heavy hammers, for the absorption of the vibration due to the blow. It is often embedded in masonry or concrete.

Anvil vise   A vise having an anvil cast as a part of the stationary jaw.

Arc of contact   In tooth gearing, the space included between those two points where the contact of a single pair of wheel teeth begins and ends.

Arc welding   Welding in which the piece to be welded is usually made the positive terminal, direct current is used, and the welding rod is the negative terminal. The work is touched with the rod and withdrawn slightly, causing an arc.

Automatic   Self-regulating or self-adjusting. A movement is automatic when it is effected without the direct intervention of the hand.

Automatic center punch   A center punch so constructed that, when pressure is applied, a spring-controlled hammer contained within the handle is released with sufficient force to cause the point to leave its mark on metal.

A.W.G. (American Wire Gauge)   Adopted as a standard for gaging the size of wires used for electrical purposes.

Awl   A small pointed tool for making holes for nails or screws.

Back gear   An arrangement of gear wheels by which the power of the driving belt is proportionately increased, as on the head of a lathe.

Back rest   A guide for supporting slender work, attached to a slide rest in a lathe when turning or grinding. When the rest follows the cutting tool, it is frequently called a "follow" or "follower rest."

Backing off   Removing metal behind the cutting edge to relieve friction in cutting, as in taps or reamers.

Backing out   The running back of a tap or die after a thread has been cut.

Ballpeen hammer   The type of hammer commonly used by machinists. One end of the head is rounded or ball-shaped for riveting or peening; the surface of the other end is flat and is used for striking a chisel, or for other such work.

Ball reamer   A hemispherical rose reamer used in finishing the recess for a ball joint.

Bar iron   Lengths of iron which are used in forging and in the shops; generally applied to that which is flat or rectangular in section.

Bars   Refers to lengths of iron or steel of rectangular sections.

Basil   The beveled edge of a drill or chisel.

Bastard   A coarse-cut file but not as rough as a first-cut.

Beam drill   Similar to a radial drill except the arm is supported at both ends, as the cross rail on a planer.

Bearing   The support or carrier for a rotating shaft.

Bearing metal   Antifriction and white metals, brass, gun metals, and the various alloys used for making or lining the bearings of journals.

Bed   That part of a lathe which supports the head stock, tail stock, and carriage.

Belt   A band or strap of leather, canvas, or other material, flexible enough to act as a transmitter of power over smooth pulleys, acting by friction only.

Belt clamp   A device for holding the two ends of a belt together while they are being connected.

Belt lacing   A narrow strip of rawhide used for lacing the ends of a belt together. Wire hooks and other types of fasteners are sometimes incorrectly spoken of as "belt lacing."

Bench assembly   The process of fitting and putting together two or more parts on a bench. The fitting may require filing, scraping, tapping, reaming, soldering, drilling, fastening with screws, and the like.

Bench dog   A peg of wood or metal inserted in a slot or hole near the end of a bench; used to prevent a piece of work from slipping. Different from bench stop.

Bench vise   The ordinary machinist's vise, either plain or swivel.

Bevel   A tool used for testing the accuracy of work cut to an angle or bevel.

Bezel   The bevel on the edge of a cutting tool.

Binary alloys   An alloy made up of only two metals.

Birmingham or Stubs wire gage   Designates Stubs soft wire sizes. Used for gaging iron wire and hot- and cold-rolled sheet steel. Differs from the Stubs steel wire gage.

Bite   Trade term for etching on metal plate.

Block chain   The type of chain used on bicycles. It consists of unit blocks connected by means of side links.

Bolt   A fastening; commonly a piece of metal with head and threaded body for the reception of a nut.

Brazing   The joining of two or more pieces of metal by means of an alloy.

Brazing metal   An alloy of 2 parts tin and 98 parts copper.

Broach   A long tool with serrated edges, which is pulled or pushed through a hole in metal to form a required shape other than round, or to enlarge the hole.

Broaching   The process of cutting or enlarging a hole in metal of a required shape usually other than round.

Brown and Sharpe taper   A very commonly used taper especially on milling-machine spindles. The rate of taper is ½ in/ft except in no. 10.

Brown and Sharpe wire gage   Also known as the American gage. Used for gaging sheets and wires of nonferrous metals such as brass, copper, and aluminum.

Buff   An appliance for polishing and finishing metallic surfaces; usually consists of a large number of muslin disks fastened together to form a polishing wheel.

Bull wheel   The large gear of a metal planing machine (planer) which drives the table. Also wheel around which a rope is wound for lifting heavy objects.

Bulldozer   A heavy forming machine for bending iron and steel.

Burnisher   Tool of hardened and polished steel for finishing metals by friction. It is held against the revolving work and gives a smooth surface by compressing the outer layer of metal.

Burr   The ragged or turned-down edge of a piece of metal resulting from grinding, cutting, or punching.

Bushing   A sleeve or liner for a bearing. Permits accurate adjustment and inexpensive repair.

Buttress thread   A screw thread which is triangular in section but which has one face at right angles to the axis of the screw, the second face only being sloped. Used in cases where excessive shock must be absorbed.

Cabbaging press   A press for compressing loose sheet-metal scrap into convenient form for handling and remelting.

Caliper   A tool for measuring the diameter of circular work.

Caliper square   A measuring instrument similar in shape to the vernier caliper but used where less accuracy is required. It consists of a fixed jaw which is an integral part of the graduated bar. The movable jaw has screw adjustment and can be locked in position.

Cam   A device mounted on a revolving shaft used for transposing rotary motion into an alternating, reciprocating, or back-and-forth motion.

Cam drive   A cam-operated mechanism through which a certain motion is made to take place in exact time or relation to some other motion, as in the camshaft of an automobile.

Cam vise   A vise whose opening and closing depend on cam action.

Cap   The top half of a journal bearing.

Cape chisel   A narrow-blade chisel for cutting channels or keyways.

Cast steel   Steel made into the desired shape by the casting process, as distinguished from other methods.

Cathead   A collar loosely fitted to a shaft and attached to it by means of headless setscrews. It is used to prevent a steady rest from marking the work.

Center drill   A short drill, used for centering work in order that it may be carried on the lathe centers. Center drills are usually made in combination with a countersink, which permits a double operation with one tool.

Center gage   A flat gage used for setting a tool for cutting threads.

Center head   A device attached to a scale or blade for use in locating the center of some round object, such as the center point on the end of a shaft preliminary to centering.

Centerless grinding   Accomplished by a machine with a high-speed grinding wheel, opposite which is a regulating wheel moving slowly in reverse direction. A rest forms a support for the work in the throat between the two wheels.

Center reamer   Center reamers, or countersinks, for centering the ends of shafts, etc., are usualy made with a 60 degree included angle. A combined center drill and reamer is now universally used.

Center square   A tool used for finding the center of a circle or of an arc of a circle. Its most frequent use is in locating the center point on the end of a shaft or cylinder to be turned.

Centering work   The process of locating the exact center of a piece of material for turning on the lathe. A hermaphrodite caliper or a combination square with a center head and a scriber may be used to center cylindrical pieces.

Centrifugal   Proceeding from the center.

Chain transmission   A means of transmitting power, useful when the distance between driver and driven shafts is too great for gearing and not sufficient for belting.

Change gears   The arrangement of lathe gears for cutting screw threads. The train of gears may be combined to give different rates of advance so that the threads of a variety of pitches may be cut.

Chase   A lengthwise groove for the reception of a part to make a joint. Also, cut threads.

Chasing threads   Cutting threads with a chaser, which usually is a flat tool containing several teeth of the desired pitch.

Chatter   Caused in machine work by lack of rigidity in the cutting tools or in machine parts.

Cherry   A form of milling cutter which is more strictly a formed reamer, for finishing out the interior of a die or some similar tool.

Chilled casting   Casting made in an iron- or steel-faced mold. The rapid cooling which takes place in such a mold tends to increase the hardness of the casting.

Chuck   A device for holding a rotating tool or work during an operation. There are many different kinds of chucks for various purposes.

Chucking   In lathe turning, the mounting of the work in the chuck.

Chucking reamer   Spiral-fluted chucking reamers with three and four grooves are employed for enlarging cored holes, etc. They are also made with oil passages through them, and in this form are adapted to operating in steel. Another style is used for finishing holes that have been roughed out.

Circular milling machine   A vertical-type continuous milling machine in which both table and cutter rotate.

Clamp   A tool for holding portions of work together.

Clamp dog   A lathe carrier consisting of two jaws and two bolts which permit clamping of the work.

Clearance   In a lathe tool, the angle between the cutting edge and vertical position of the work. The clearance must be not less than 3 degrees and in most cases not more than 10 degrees. Also, the amount of space, open or free, between adjacent parts.

Climb cut   The cutting of metal, in a milling machine, in which the table moves in the direction of the rotation of the cutting tool.

Cold-rolled steel   Steel made by either the open-hearth or the Bessemer process. The carbon content runs from 0.12 to 0.20 percent. This steel is marketed with a bright, smooth surface and is made quite accurate to size so that for many purposes no machining is necessary. It may be case-hardened, but it will not temper.

Collapsing tap   This type of tap is used in machine tapping. The principle is similar to that of self-opening dies except that the action is reversed. Such taps permit more rapid tapping and do not mar the thread.

Collar   A ring formed on a shaft by forging in the solid, or by being made as a separate piece, bored and turned and held in place by a setscrew or pin.

Collar oiling   Is used on high-speed installations. A collar which runs in a reservoir of oil in the bearing is either turned on the journal or is attached to it. Wipers are often used in the upper half of the bearing to scrape oil off the collar and spread it on the journal.

Collar stud   A stud threaded at one end and having a short shaft or spindle at the other, the two separated by a collar which is an integral part of the stud. Used for carrying gears, levers, etc.

Collet   A clamping ring or holding device; in the shop the term is freely applied to sockets for tapered-shank drill and to reducing sleeves and bushings of various types.

Combination chuck   A universal chuck which may also have independent jaw action.

Combination die   A die so constructed that it both cuts the blank and draws the piece to required shape. A combination of blanking and drawing dies.

Combination drill and countersink   Used for centering work to be held between the centers of a lathe.

Compound rest   A secondary slide, tool rest, and swivel plate superimposed on the cross slide that can be hand fed independently of the lathe cross feed. It is of particular advantage in turning to a desired angle.

Compound sliding table   A machine-tool table having at least two movements, one longitudinal and the other transverse.

Cone   A piece like the circular segment of a cone.

Cone mandrel   A type of mandrel on which work is centered by the clamping action of two cones against the work. The mandrel has a shoulder to prevent one cone from sliding, and threads to permit locking of the second cone.

Cone pulley   A stepped pulley, one having two or more faces of different diameters. Used in pairs, the large end of one being opposite the small end of the other, so that a shifting of the belt will give a change of speed.

Cone pulley lathe   A special lathe used for turning cone pulleys.

Corner rounding cutter   Used on milling machines for finishing rounded edges.

Corundum   An extremely hard aluminum oxide used as an abrasive.

Cotter mill   A milling cutter used for cutting key seats, slots, and grooves.

Counterbore   A cutting tool fitted with a pilot or leader to guide and center the cutting edges. Used, for example, to enlarge the beginning of a drilled hole so that the head of a screw will be flush with the metal surface.

Countershaft   The intermediate shaft between line shaft and machine.

Countersink   To recess a hole conically for the head of a screw or rivet.

Crest of screw thread   The top surface joining the two sides of a thread.

Critical temperature   The temperature (varying for different steels) at which certain changes take place in the chemical composition of the steel during heating and cooling.

Crossbelt   A belt changed to run from the top of one pulley to the bottom of another to produce a reversal of direction.

Cross feed   A transverse feed. In a lathe, that which ususally operates at right angles to the axis of the work. In a planer or shaper, that feed which carries the tool across the work.

Cross slide   The horizontal slide or bridge which carries the tool box in a planing machine. The slide which operates on the saddle of a lathe carriage to provide for transverse feed. It also supports the compound rest.

Cross valve   A valve fitted on a transverse pipe so as to open communication at will between two parallel lines of piping. Much used in oil- and water-pumping arrangements.

Crossing file   A taper file with a section like two half-rounds with flat faces back to back.

Crown pully   A pulley whose diameter is greater at the middle than at the edges of its face. This crown tends to prevent the belt from running off the pulley provided the belt is not slipping.

Cup center   The center which is used in the tail stock of a wood-turning lathe; also called a "dead center."

Curling die   A die used for making a curled edge at the top of any cuplike piece drawn from sheet metal.

Dead center   The center which is fitted into the tail stock of a lathe and does not rotate with the work.

Dead level   An emphatic expression in the sense of absolutely level.

Dead-smooth file   The finest cut file made.

Depth gage   A gage used by metalworkers for testing the depth of holes and recessed portions.

Depth micrometer   A precision gage with micrometer adjustment, used to determine the depth of holes, slots, and counterbores, and the distance from one surface to a lower level.

Diamond-point chisel   Similar to a cape chisel except that its point is shaped for the cutting of sharp-bottomed grooves.

Die   An internal screw used for cutting an outside thread.

Die chase   A threaded section of a screw-cutting die.

Diehead   The device which carries the threading dies in a screw-cutting machine.

Differential indexing   In differential indexing on the milling machine, the index plate is connected to the spindel on which the work is mounted, through change gears. Both plate and gears, therefore, revolve to produce the proper indexing.

Direct indexing   Direct indexing on the milling machine is done in such a manner that the spindle and the work mounted thereon revolve through identical angles with the indexing plate.

Dividing head   A mechanical device which provides for the dividing of the circumference or perimeter of a piece of work into equal parts or spaces, as in the spacing of gear teeth.

Doctor   Local term for an adjuster or adapter that allows chucks from one lathe to be used on another. The term "dutchman" is sometimes used in the same way.

Dog   The carrier of a lathe. One of the jaws of a chuck.

Double-action press   A press for handling two operations for each revolution of the press. It carries two rams, one inside the other, so actuated that one motion immediately follows the other.

Double-geared   A lathe or drill press equipped with an ordinary back gear is said to be double-geared.

Draw chisel   A pointed cold chisel used for shifting the center for a hole to be drilled, in cases where the drilling has been started but the drill has run, i.e., is not drilling in the exact position desired.

Draw chuck   Used on small, accurate work. It operates by a longitudinal motion in a taper bearing.

Draw filing   A metal polishing operation, with a single cut file, for the purpose of removing finish marks, scratches, and nicks from the face or edges of a metal object.

Drift pin   A round tapered pin driven into rivet holes when they are not in perfect alignment. In some cases the holes may be distorted sufficiently to permit the setting of the rivet.

Drift punch   A tool used to align rivet or bolt holes in adjacent parts so that they will coincide, or to drive out pins or rivets.

Drill gage A flat steel plate drilled with holes of different sizes and properly marked so that the size of a drill may be easily determined by fitting it to the plate.

Drill-grinding gage Used for checking the length and angle of the cutting lips of the drill.

Drill press A geared, automatic-feeding machine tool used for drilling holes in metal.

Drill rod A 90- to 100-point carbon, ground and polished tool steel finished to a limit of .0005 in.

Drill spindle The vertical spindle of a drilling machine which carries the drill and revolves, and through whose vertical movement the feed is operated.

Drill vise A vise used on the table of a drill press for holding work to be drilled.

Drilled hole A hole finished by means of a drill.

Drop-feed oiler Sight-feed oil cup equipped with a needle valve which can be adjusted to control the flow of oil.

Drop hanger A shafting support attached to a ceiling or underside of a beam.

Eccentric flute A reamer so milled that the concave channels are unevenly spaced to obviate chattering, but placed exactly opposite to permit diameter measurement with a micrometer.

Eccentric turning Lathe turning work which is not concentric with the axis.

End mill Milling cutter usually having tapered shanks, for direct fitting to spindles or sockets. The cutting portion is cylindrical in shape, so made that it can cut both on sides and end.

Engine lathe A lathe having cross slide, compound rest, lead screw, and power feed; equipped with changer gears and back gearing.

Expanding mandrel A form of lathe mandrel or arbor which is made to expand within the work.

Expansion bit A boring bit having a cutter or cutters arranged to permit radial adjustment, to enable one tool to bore holes of different diameters.

Expansion reamer   A reamer which admits of a limited amount of adjustment for size. Such adjustment is usually through wedge action controlled by a screw.

External thread   The thread on the outside of a screw or bolt.

Face   The face of a casting is that surface which is turned or polished. the face of a gear wheel indicates the breadth of the teeth; of a belt pulley, the breadth of the rim.

Faceplate or face chuck   A circular plate for attachment to the spindle in the headstock of a lathe. Work may be clamped or bolted to it. The slots engage the tail of the lathe dog.

Fault   Any flaw or imperfection.

Feather or sunk key   A parallel key partly sunk into a recess in a shaft so as to form an integral part of the shaft. The keyway in the wheel or clutch carried on the shaft is made large enough to permit these parts to slide longitudinally on the shaft.

Feed   In machine-tool operations, the rate of tool travel across a surface from which material is being removed.

Feed gears   Gears used to drive the feed rod and control the rate of feed.

Feed mechanism   An arrangement of gears, screws, and other devices for controlling the feed of the tool to the work or the work to the tool.

Feed screw   The screw whose rotation imparts a measurable amount of feed to the cutting tool of a machine.

Feeler   Gauge for determining the size of a piece of work, the accuracy of the test depending on the sense of touch.

Files, kinds of   Many kinds of files are made for special purposes and known by the name of the purpose for which they are intended. In general, however, files are named for the manner in which they are cut, as bastard, rough-cut, second-cut, smooth, etc.

Fine pitch   A relative term. A gear with small teeth, or a screw with a comparatively high number of threads per inch, is said to be of fine pitch.

Finishing cut   A cut usually of small depth and fine feed for imparting a smooth surface and bringing the work to the desired size.

Finishing tool   Commonly applied to the metalworker's cutting tool whose cutting edges are broad and straight for removal of the ridges left upon work by the roughing tools.

Firm-joint caliper   Ordinary calipers whose two legs are attached by a large firm joint instead of a rivet.

Fishtail cutter   A type of cutter for cutting grooves or seats in shafts; suitable for light cut and feed.

Fit (thread)   Loose fit (class 1), standard for tapped holes in numbered sizes only; free fit (class 2), generally used on work of average quality—also called "medium fit, regular"; medium fit (class 3), used on the better grade of screw-thread work; close fit (class 4), used on screw-thread work where close fit is necessary. Shrink fit, force fit, and drive fit are common designations of fits used in the nonthread assembly of nonthreaded parts.

Flange   A rib or offset on a casting. The circular faces of couplings or of pipe fittings. The turned edge of a metal shape or plate, which resists bending strain.

Flanged pulley   A pulley having a flange or increase in diameter on one edge of its face is a single-flanged pulley; with an increase in diameter on both edges, it is a double-flanged pulley. The flange prevents the belt from slipping off.

Floating tool   A tool so secured in its holder that it may be guided in its operation by the piece on which it works.

Flute   The concave channel in a reamer, tap, or drill.

Fluting cutter   Used on a milling machine for fluting taps, reamers, etc.

Fly cutters   Cutters set in a cutter block or chuck held in a lathe or other machine, and used for shaping the ends of metal rods or for other formed work.

Flywheel   A heavy wheel used in machinery where reciprocal motion is converted into circular motion. It aids in maintaining uniformity of motion.

Follower   A wheel which is driven by another wheel. Also, the roller which operates against a cam face, or the roller and the arm to which it is attached.

Follower rest   A support for lathe work, attached to the carriage. It follows or is opposite the cutting tool, thus preventing the work from springing away from the tool.

Footstep or footstep bearing   A bearing used at the lower end of a vertical shaft or spindle to carry the end thrust.

Forging press   A machine used to exert the pressure needed in die forging.

Forked center   A center with taper or straight shank and head for holding cylindrical objects in position during drilling and other operations.

Fox lathe   A brass worker's lathe which has a chasing bar or "fox" for thread cutting.

Fracture   To break apart; to separate the continuous parts of an object by sudden shock or by excessive strain.

Full thread   A screw thread which is cut clean and sharply to its proper depth.

Gang mills   A series of milling cutters arranged on the same mandrel to increase production; may be used to machine two or more surfaces at one time.

Gap bed   A lathe bed having a recessed portion in front of and below the head stock, to receive work larger in diameter than the lathe would otherwise accommodate. The bed is strengthened below the gap. The gap, when not in use, is filled with a gap bridge.

Gashing   The rough cutting of machine parts, particularly the teeth in bevel gears.

Gear   A very general term applied to toothed wheels, valve motions, pump work, lifting tackle, ropes, etc. The point or portion of the tooth of a gear wheel lying outside the pitch circle is the addendum; that portion between the pitch and the root lines is the dedendum.

Gear cutters   Circular cutter of hardened steel whose section is that of the tooth spaces which they are intended to cut.

Geared chuck   A form of universal chuck.

Geared head   A head stock equipped with back gear.

Gear-tooth caliper A vernier-type caliper used for measuring the depth and thickness of gear teeth on the pitch line.

Gib (1) A thin piece of steel used as an adjusting strip to bring about a perfect sliding fit in machine parts, e.g., in the cross slide of a lathe. Adjustment is usually secured by the pressure of setscrews against the gib. (2) That portion of a gib and cotter arrangement used in the strap end of a connecting rod to keep it from spreading. It is a flat piece of steel with hook ends.

Grind finish A finish imparted by an abrasive wheel. A great deal of the work is now finished by grinding that was formerly tool finished.

Grinding allowance The amount of material left on a piece of work to allow for a finish by grinding. On fine work .003 to .007 in is sufficient, while on heavy work it may be $\frac{1}{64}$ or $\frac{1}{32}$ in.

Grinding in Bringing to a perfect fit, by means of an abrasive, parts which are to operate together. This may be accomplished either by a hand or machine process.

Gripe A name occasionally applied to a machine clamp.

Grooving The cutting of a groove or channel.

Guide bearings Bearings which consist of a channel or groove in which parts slide, e.g., crosshead bearings.

Gun lathe A large lathe used for turning and boring cannons.

Hand feed The feeding by hand of cutting tools of machines of various kinds.

Hand miller A small milling machine with the feed operated by hand. It is adapted only to light work.

Handwheel Any wheel operated by hand, usually to secure an adjustment.

Head stock The fixed head of a lathe which carries the faceplate or chuck.

Helical angle The angle which any portion of a helix or screw makes with a line drawn at right angles to its axis.

Helical gear A tooth gear in which the wheel teeth, instead of being at right angles with their faces, are set at some other angle therewith. Often incorrectly called spiral gear. It may be used to transmit power between (1) parallel shafts, (2) shafts at right angles and not intersecting, or (3) shafts inclined at any angle and not intersecting. It gives greater strength and smoother operation, but develops considerable end thrust.

Helical spring A compression-type spring shaped like the frustum of a cone.

Helix angle The angle made by the helix of a thread at the pitch diameter with a plane perpendicular to the axis.

Hermaphrodite caliper A caliper in which one leg is pointed as in a pair of dividers, the other being slightly hooked as in the ordinary outside caliper.

Hex head A common shop expression referring to screws and bolts with hexagonal heads.

High spots (1) Spots to be taken down by scraping or grinding in order to secure an absolutely plane surface. (2) Area of a plane extending above a true plane.

Hogging A term frequently used in machine-shop practice referring to very heavy cuts taken on machine tools.

Hook spanner Use to turn round nuts notched on the periphery.

Horning press Used for closing down side seams on sheet-metal receptacles, such as buckets.

Hotbox A bearing which, from a poor fit or lack of lubrication, has become overheated.

Hypoid An abbreviation of hyperboloidal. The term is applied to a special type of spiral-bevel gear tooth.

Index To space work off in equal divisions by means of an index head or similar device.

Index head or dividing head A mechanical device, usually attached to a machine, used for the purpose of spacing circular work into equal divisions.

Indexing Dividing a circle into regular space for purposes of milling, fluting, or gear cutting.

Individual drive A direct motor drive dedicated to a unit as opposed to a single motor drive that supplies power to multiple units through counter-shafts from a line shaft.

Inserted tooth cutter A milling cutter having inserted teeth which are held in position by various methods. Inserted teeth are commonly used when the cutter is 6 in or more in diameter.

Inside calipers A calipers with the points at the end of the legs turned outward instead of inward, so that it may be used for gaging inside diameters.

Internal thread A thread cut inside a part such as a nut.

Jarno taper This taper of 0.6 in/ft is used by a number of manufacturers for taper pins, sockets, and shanks on machine tools.

Jig A device which holds and locates a piece of work and guides the tools which operate upon it.

Jig boring A boring operation on a piece of work which is held in a jig, the tool being guided by some portion of the jig. The term boring indicates that the work is performed with a special boring tool, not with a drill.

Jig bushing Hardened-steel bushing inserted in the face of a jig to serve as a guide for drills.

Jigger A mechanism which operates with quick up-and-down motion; a jolting device.

Kennedy key Consists of two square-bodied keys so placed that diagonal corners intersect the circumference of the bore. Used for heavy work.

Key A wedge-shaped strip of iron or steel that prevents wheels from slipping on their axles. Keys are of various kinds and shapes.

Key seat The recessed groove or space, in either shaft or hub, made to receive a key.

Land Space between flutes or grooves in drills, taps, reamers, and other tools.

Lap An accurately finished tool with a surface charged with an abrasive substance.

Lapping The finishing of external or internal surfaces either by hand or by machine.

Lathe   A machine used for the production of circular work.

Lathe, engine   The ordinary form of lathe with lead screw, power feed, etc.

Lathe, gap   A lathe with a gap or cutout in front of the head stock to increase the swing for faceplate work.

Lathe bed   The longitudinal supports for the head stock, tail stock, and the slide rest of a lathe.

Lathe center grinder   A grinding device which can be attached to a lathe and used for grinding centers.

Lathe chuck   Form of holding device, attached to a lathe spindle which grips the work while it is being operated on.

Lathe dog   A carrier; an attachment which may be fastened to lathe work, and has a projecting tail to engage in a slot or hole in the faceplate.

Lathe shears   The machined top of the lathe bed.

Lathe tool   Also called "cutting tool." Used for removing excess stock from metal worked in a lathe. The commonly used lathe tools are: side tool, diamond point, bullnose, inside boring, threading, cutting off. Usually made of high-speed or carbon steel, and ground with wet emery grinder.

Lathe work   The work commonly done in the lathe; includes almost all branches of turning and boring.

Laying out   The setting off or marking out of work to full size.

Layout bench or table   A bench with a level metal top on which work can be laid out.

Lead   The distance a screw advances when given a single complete turn. In a single thread the lead is equal to the pitch. In a multiple-thread screw the lead equals the multiple times the pitch; that is, in a triple thread the lead is 3 times the pitch.

Lead hammer   A hammer with a head made of lead; used in place of a steel hammer to avoid bruising of parts. Also, a hammer made of lead and sometimes used for the same purpose as a copper hammer, for hammering against harder metal which would be bruised if a steel hammer were used.

Lead hole   A hole drilled in a piece of metal to facilitate the drilling of a larger hole, or to assist in centering a drill upon an inclined surface.

Lead screw The screw which runs longitudinally in front of the bed of a screw-cutting lathe.

Left-hand thread A screw thread so cut that the bolt, screw, or nut has to be turned in a counterclockwise motion to engage or tighten it.

Left-hand tool Side tool ground to an angle on the right-hand side and which, therefore, cuts from left to right.

Let-in A shop term to signify the sinking in of one portion of metal into another.

Lever A rigid bar turning upon an axis or fulcrum.

Light cut In metalwork, a cut is said to be light when the shavings removed are thin and narrow.

Limit gage To permit interchangeability, a limit of variation is permitted on each side of the correct dimension. Gages are made to these limits, and used to test the work.

Limits of tolerance Refers to limits of accuracy, oversize, or undersize, within which a part being made must be kept to be acceptable.

Line shaft A run of shafting which consists of several lengths coupled together. It may or may not be a main-line shaft.

Live center The center in the revolving spindle of a lathe or similar machine. It is highly important that this should run true or it will cause the work to move in an eccentric path.

Live spindle The revolving spindle in the head stock of a lathe, as opposed to the dead spindle of the tail stock.

Lockpin Any pin or plug inserted in a part to prevent play or motion in the part so fastened.

Lowenherz thread A German thread with flats at top and root, similar in appearance to U.S.S. thread but having a different included angle.

Lubricant Used to reduce friction and exert a cooling action on tool and material that is being cut. The lubricant used for this purpose also removes chips and imparts a smoother finish.

Machine A device for transforming or transferring energy.

Machine drilling The drilling of work under a power-driven machine.

Machinery steel   An open-hearth steel with 0.15 to 0.25 percent carbon content. The term is rather general in its use and is frequently applied to any mild steel which cannot be tempered but may be case-hardened.

Machine tool   The name given to any machine of that class which, taken as a group, can reproduce themselves, such as the lathe, drilling machine, planer, milling machine, etc. No other class of machines can be used to build other machines, and because of this, machine tools are known as the "master tools of industry."

Machining   The operations performed by machines on metalwork.

Machinist   One who operates machine tools is classed as a machine hand. One who makes and fits parts by hand is called a bench hand. One who assembles machines is often referred to as a floor hand.

Mandrel   A shaft or spindle on which an object may be fixed for rotation; e.g., when a piece is to be turned in a lathe, it may be carried on a "mandrel" which is supported by the lathe centers. The terms mandrel and arbor are often used interchangeably.

Manual   Relating to handwork. That which is done by hand.

Master taper   A standard taper, either inside or outside, by which other tapers are tested.

Medium fit   Used for sliding and running fits on machinery where greater accuracy is necessary than would obtain with "free fit."

Medium force fit   Such fits are effected through considerable pressure.

Metal spinning   The process by which light articles in the malleable metals are made to assume circular and molded shapes, through pressure applied to them while in rapid rotation in the lathe.

Mill   To machine with rotating toothed cutters on a milling machine.

Mill file   A single-cut file made in any cut from rough to dead smooth. Used for lathe work, draw filing, etc.

Milling cutters   All those various types of rotary cutters designed for use on a milling machine.

Milling machine   A machine in which the operating tool is a revolving cutter. It has a table for carrying work and moving it so as to feed against the cutter.

Milling machine, universal A milling machine in which the work table and feeds are so arranged that all classes of plane, circular, helical, index, or other milling may be done. It is equipped with index centers, chuck, etc.

Milling machine, vertical It differs from the horizontal machine mainly in having a vertical spindle for carrying the cutter.

Miter gear A bevel gear whose pitch cone is placed at an exact angle of 45 degrees with its axis. Pairs of miter wheels working together are always of equal diameter, pitch, and number of teeth.

Morse taper Standard taper from 0 to 7 for fitting the shanks of drills and other tools to machine spindles.

Multipart bearing A bearing consisting primarily of three or more sections in contact with the journal, and arranged to give the least interference with an oil film. These sections are contained in a housing. Bearings of this type are used on heavy-duty installations.

Multiple-thread screw A screw with several helixes winding around its body. Used to impart more rapid motion than could be obtained by a single-threaded screw.

National Coarse thread The screw thread of common use, formerly known as the United States Standard thread.

National Fine thread Same form as National Coarse threads but of finer pitch. Much used in automobile work.

Necking tool Tool for turning a groove or neck in a piece of work.

Nipple chuck A rapid-adjusting machine chuck, used in production work for holding pipe while nipples are cut and threaded.

Nose The business end of tools or things. The threaded end of a lathe or milling-machine spindle, or the end of a hog-nose drill or similar tool.

Nut arbor or nut mandrel An arbor on which nuts are finished to shape.

Oddleg caliper Calipers with moderately curved legs, both curved in the same direction, used for measuring shoulder distances, etc.

Off-center A term applied to a part which does not run true.

Oil groove Small, semicircular groove, cut in the internal face of a bearing and on the sliding surface of machinery, for the distribution of oil for lubricating purposes.

Oil hardening   The hardening of steel by quenching it in oil instead of water.

Open washer   Also called "slip washer." A washer partly cut away so that it may be slipped around a bolt without entirely removing the nut.

Operator   One who manipulates a machine or controls the working thereof.

Out-of-gear   When the teeth of gear wheels, which usually mesh together, are disengaged or when the components of the driving mechanism of a machine are disconnected from the rest of the machine by clutch or other means, they are said to be out of gear.

Output   Amount of energy delivered by the source of generation to an external device.

Overall   A common term in shopwork meaning an outermost or total dimension.

Pack hardening   Treating steel with some carbonaceous material and quenching it in oil. The terms "pack hardening" and "case hardening" are often used interchangeably.

Peripheral speed   The speed, usually registered in feet per minute, of the circumference of a part such as a wheel or shaft.

Pillar file   Used on narrow work such as cutting grooves. Same general shape as a hand file except that it is not as wide and is obtainable in any cut.

Pin spanner   Used to turn round nuts having holes in the periphery to receive the spanner pin.

Pin vise   A small hand vise with a notch in each jaw for gripping wire or round objects.

Pin wrench   Used on round nuts which have two holes in their face to receive the pins of the wrench.

Pipe thread   The thread used on pipe and tubing, cut on a taper of ¾ in/ft, which ensures a thoroughly tight joint.

Pitch of a screw   The distance from a point on a screw thread to a corresponding point on the next turn. In a single thread the pitch is the amount of advance in one revolution.

Pitch of gears   Refers to the size of gear teeth.

Plain turning   Straight or cylindrical turning.

Planchet   Blank piece of metal punched out of a sheet before being finished by further work.

Plane   Level, flat, even.

Planer   A metalworking machine for producing plane surfaces.

Planometer   A name sometimes given a surface plate.

Platen   A flat working surface for laying out or assembling metalwork. Also, the movable table of a planer or similar machine.

Play   The motion between poorly fitted or worn parts.

Plug gage   A very accurately made plug for testing the size of holes or internal diameters in machine work.

Poppet   The head stock of a lathe. A lathe center. The term is not generally used.

Positive feed   When the feed motion is communicated directly by means of gears, without friction clutches or belts.

Power feed   The automatic feed of a lathe, planer, screw cutter, or other machine.

Precision grinding   Machine grinding in which the tolerances are exceedingly close.

Precision lathe   A small bench lathe used for very accurate work.

Press fit   A fitting together of parts by pressure; slightly tighter than a sliding fit.

Prick punch   A small center punch. Also known as a layout punch.

Profiling machine   A type of milling machine in which the cutter can be made to follow a profile or pattern. A very valuable machine for certain classes of work.

Pulley lathe   A lathe used for turning either a straight or crowned face on pulleys.

Pulley tap   A tap with a very long shank, used for tapping setscrew holes in the hubs of pulleys.

Pull pin   A device for throwing mechanical parts in or out of gear, or for readily shifting in or away from a fixed relative position.

Quadruple thread   A thread in which there are four distinct helices, making the lead 4 times the pitch. A quadruple thread is usually of the square or Acme type.

Quality   Degree of goodness.

Quartile   Refers to a quarter part of.

Quaternary steel   That class of alloy steel which consists of iron, carbon, and two other special elements.

Quenching   The dipping of heated steel into water, oil, or other bath, to impart necessary hardness.

Quick return   A term applied to shapers, planers, and other metalworking machines which are so constructed that the return stroke is much more rapid than the forward or cutting stroke.

Quill   A hollow shaft or spindle.

Quill gear   A gear or pinion cut on a quill or sleeve.

Radial arm   The movable cantilever which supports the drilling saddle in a radial drilling machine.

Radial drilling machine   A heavy drilling machine, so constructed that the position of the drill can be adjusted to the work without moving the latter.

Rake   The amount of set on a cutting tool.

Rake angle   In a milling cutter, the angle measured between the face of the tooth and a radial line in a diametral plane. In a lathe tool the angle between the upper side of the cutting face of the tool and horizontal face of the tool holder or shank.

Ram   The movable part of a shaper which carries the tool.

Ratchet   A gear with triangular-shaped teeth, adapted to be engaged by a pawl, which either imparts intermittent motion to the ratchet or locks it against backward movement when operated otherwise.

Ratchet bar   A straight bar with teeth like those of a ratchet wheel to receive the thrust of a pawl.

Ratchet drill A hand drill in which a lever at one end of a drill holder is revolved by a ratchet wheel and pawl.

Ratcheting end wrench A nonslip end wrench that permits working in close quarters; grip is released by a slight backward movement and a new hold is had automatically without removing the wrench.

Rate of speed In machine work rate of speed may be expressed in revolutions per minute (rpm) or in feet per minute.

Ream To smooth the surface of a hole and finish it to size with a reamer as for a running fit.

Reamer A tool with cutting edges, square or fluted, used for finishing drilled holes.

Reaming The process of smoothing the surface of holes with a reamer.

Rechucking The resetting of a piece of work in a chuck for further operations.

Reed taper One of the several standard tapers used on lathe spindles.

Relieving Removing of material back of the cutting edge of a tool to reduce friction, as on milling cutters.

Revolution The act of revolving, as the turning in a complete circuit of a body on its axis. Usually distinguished from rotation, which may mean a revolution or a part of a revolution, the term revolution is applied to continuous motion such as that of a shaft.

Revolutions per minute An expression for the rotational speed of machines. Abbreviated as rpm.

Riffler A small rasp or file, usually curved, used for filing inside surfaces or for enlarging holes.

Ring gage A gauge in the shape of a ring used for checking external diameters.

Ring oiler A bearing in which the journal carries a loose-fitting ring, the lower half being immersed in oil. As the shaft rotates the ring is given a circular motion, picking up oil and distributing it to the bearing.

Root That surface of a thread which lies between adjacent threads.

Root diameter The diameter at the bottom of a thread.

Rose bit   A solid, cylindrical parallel boring tool used for finishing drilled holes.

Rose reamer   A heavy-duty machine reamer, designed so that cutting is done by its end, which is beveled, instead of by its sides.

Rotary cutter   A cutter which rotates with the spindle to which it is attached, to produce a cut on the work with which it comes in contact.

Rough cut   Usually the first or heavy cut taken in preparation for the finish cut.

Roughing tool   The ordinary tool used by machinists for removing the outer skin and generally for heavy cuts on cast iron, wrought iron, and steel.

Round-nose tool   A type of tool used for roughing cuts and for turning fillets.

Round-point chisel   Used for cutting oil grooves, etc.

Running fit   Refers to the fitting together of parts with just sufficient clearance to permit freedom of motion.

Saddle   The base of a slide rest which lies on the lathe bed. Also the sliding plate which carries the drill spindle and gear wheels of a radial drill, or the crossbar which carries the tool head on a planer or boring mill.

Scrape   To finish a surface or to fit a bearing by the use of a hand tool called a scraper.

Scraped joint   A joint, such as a bearing, brought to a perfect fit by means of a scraper.

Scraper   A tool used by metalworkers for fitting bearings, and for truing surfaces. They are of a great variety of shapes, depending on the work to be done.

Screw-cutting lathe   A lathe adapted to thread cutting; equipped with lead screw and change gears.

Screw pitch gage   A gage usually made up of many leaves, the edge of each being cut to a thread of indicated size. Used for determining the number of threads per inch on a given screw, bolt, or nut.

Screw plate   Originally a steel plate having holes of different sizes which are internally threaded for making screw threads by forcing up the metal. The present use of the term refers to a stock and a halved or solid die.

Screw-slot cutter A milling cutter used for slotting screw heads.

Screw stock A shop term for soft steel used for small screws and parts made on screw machines.

Screw threads Projections left by cutting a helical groove on a cylinder. Threads may be internal or external.

Self-centering or bell center punch A center punch which slides in a bell-mouthed casing. A fairly accurate centering is secured by placing the bell mouth over the end of the piece to be marked, then tapping the punch with a hammer.

Self-locking setscrews There are several patented types of setscrews which resist the loosening effect due to vibration. One type has left-hand spiral knurling at the point; another has a deeply slotted head, the slot being spread to offer resistance.

Self-opening die A die which opens automatically to permit its removal after the thread-cutting operation is finished.

Sellers screw thread The United States Standard thread of type, with flats at top and bottom equal to one eighth of the pitch. The included angle is 60 degrees.

Sellers taper This taper is .75 in/ft. It is not as much used as the Brown and Sharpe or the Morse taper.

Serial taps A set of taps usually numbered 1, 2, and 3. No. 1 is a tapered tap, no. 2 is slightly tapered near the end, and no. 3 is a full-threaded tap, called a bottoming tap.

Set at zero To set to a given point from which other adjustments can be made. In lathe work when the gage line on the tail stock is set at zero, the live and dead centers are in alignment.

Setover Transverse movement of a lathe tail stock center on its base to obtain a taper on a turned piece.

Setscrew A plain screw having a head of square or other shape used for tightening and for locking adjustable parts in position. Setscrews are usually case-hardened.

Shank   That part of a tool by which it is connected to its handle or socket.

Shaper   A metalworking machine on which the work is fastened to a table or "knee." The tool is moved back and forth over it by means of a sliding ram.

Shedder   A sort of stripper which ejects the blanks from a compound die.

Shell drill   A hollow drill carried on an arbor when in use. It is used for enlarging holes through which a drill has been passed.

Shell-end cutter   A heavy-duty cutter designed for use on a tapered arbor to which it is attached by a nut which fits into a recess at the outer end. The back end of the cutter is slotted to receive a projection on the arbor, thus giving additional driving strength.

Shell reamer   A hollow reamer which is mounted on an arbor when in use.

Shop work   Mechanical work performed in a shop.

Side-milling cutter   Cutter of comparatively narrow face which cuts both on the periphery and the sides. When two or more cutters are set up on an arbor, they are called "straddle mills."

Side rake   The amount of transverse slope away from the cutting edge on the top face of lathes, planers, shaper tools, etc.

Single-thread screw   A screw having a single helix or thread. Its pitch and lead are equal.

Skin   The thin film of hard metal on the surface of castings.

Slabbing cutter   A wide-faced milling cutter with nicked teeth to permit an easier cut than would be possible with a plain-toothed cutter.

Slide caliper   A pocket caliper consisting of a graduated bar which slides in a retaining piece.

Slide rest   On a lathe, the parts above the saddle which support the compound rest.

Slitting saw for metal   Thin milling cutters used for splitting bushings, etc.

Slotting machine   A machine used for shaping metals and cutting mortises, the arm which carries the cutters moving in a vertical direction.

Soft iron Iron which can be worked with ordinary cutting tools or which can be readily abraded with files. It is gray, as distinguished from the harder cast iron which is lighter in color.

Soft solder Solder such as is used for tin plate and other metal sheets. The composition varies from "half and half"—half lead and half tin—to 90 parts tin and 10 of lead. A very small percentage of antimony is often added.

Soldering The uniting under proper heat of pieces of metal by means of a dissimilar metal or alloy.

Soldering copper A tool, also called soldering iron, used for applying heat to melt the solder and heat up the metals that are to be joined by soldering.

Solid bearing A one-piece rigid bearing. Its use is limited as it must be slipped over the end of the shaft which it is to support. When solid bearings are pressed into the parts to which they are applied they are called bushings.

Space washer A washer used for spacing rather than to provide bearing for a nut.

Speed lathe A metalworker's lathe, not equipped with mechanical feed.

Spindle A rotating rod or arbor, either hollow or solid.

Spiral coupling A type of jaw coupling which remains engaged only when rotating in one direction.

Spline An arbor fitted with a key or keyway.

Spot face To finish a round spot on a surface, usually around a drilled hole, to give a good bearing for a screw or bolt head.

Spotting the center Spotting work with a center punch, then starting a hole with a much smaller drill than the one with which the hole is to be finished.

Spotting tool Used to spot a center and to face the end of stock. Also called "centering and facing tool."

Spring chuck or spring collet A type of chuck used on screw machines. It consists of a sleeve slotted through a portion of its length, and is closed on the work by being drawn or pressed into a conical cap into which it fits. When released, it springs open sufficiently to free the stock.

Steady rest   A rest attached to the ways of a lathe for supporting long, slender work while it is being machined.

Step block or step bearing   A bearing which takes the end thrust of vertical shafts.

Stock   A general term referring to material to be worked on, especially shafting and bars such as round stock and bar stock.

Stocks and dies   The threaded blocks or dies together with levers with which they are operated; used for cutting male threads.

Stop   A piece attached to some part to prevent motion beyond a certain point, as on a machine in shopwork.

Straightedge   A parallel, straight strip of wood or metal used for gaging the linear accuracy of work.

Straight-shank milling cutter   Used in a profiling machine for die work, routing, etc.

Strip   To break, tear, or strip off the threads of bolt or nut.

Swing   The swing of a lathe signifies the size of the piece of work which can be turned in it.

S wrench   A wrench shaped like the letter S having either fixed or adjustable openings.

Tail stock   The movable head of a lathe as distinguished from the head stock, which is fixed.

Tail stock spindle   The sleeve or spindle which carries the dead center in a lathe tail stock.

Taking up   Making adjustment for wear, as in machinery.

Tang   The shank of a cutting tool, or that portion which is driven into the handle.

Tap   The process of cutting threads with a tap, fluted, threaded tool for cutting female or inside threads.

Tap bolt   A bolt usually threaded for its entire length. It is finished only on the point and the underside of the head. Tap bolts are made with both square and hexagon heads.

Taper A gradual and uniform decrease in size, as a tapered socket, a tapered shaft, a tapered shank.

Taper attachment An adjustable mechanism attached to a lathe which permits accurate taper turning.

Tapered-shank drill A drill, twist or otherwise, whose shank is tapered, for use with an ordinary drill spindle or socket.

Tapered spindle A spindle containing a tapered recess for the reception of a center or a tapered-shank tool.

Taper gage A gage for testing the accuracy of tapers, either inside or outside.

Taper per foot A way of expressing the amount of taper; for example, the Jarno taper is .6 in/ft, and Brown and Sharpe is .5 in/ft, except for no. 10.

Taper pin Made of round stock, used for fastening some part to a shaft. It is graded in size by numbers from 1 to 10. No. 1 is .156 in in diameter at the large end, and is from ¾ to 1 in long. No. 10 is .706 in in diameter at the large end and 1½ to 6 in long.

Taper reamer Reamer of the ordinary fluted type for reaming tapered holes, such as a reamer used to prepare a hole for a tapered pin.

Taper tap A tap tapered in the direction of its length, in order to afford ease in cutting when commencing a screw thread in a drilled hole.

Taper turning Lathe turning accomplished by setting over the tail stock or by using of a taper attachment.

Tap, hob, Sellers A long tap, threaded only along the central portion of its length and containing many flutes; it is used for threading dies and "chasers."

Tapped faceplate A faceplate having tapped holes instead of or in addition to slots.

Tapper tap Special tap used for tapping nuts in tapping machines.

Tapping machine A machine frequently used in production work on small parts. It has a forward motion for running a tap into a hole and a reverse motion for backing it out.

Tap wrench   The double-armed lever with which a tap is gripped and operated during the process of tapping holes.

T bolt   A bolt shaped like the letter T, the head being a transverse piece, which fits into recessed undercut T slots, as on the table of a milling machine or planing machine.

Thread-cutting screws   The entering end of such screws contains a slot with serrated cutting edge which makes the use of a tap unnecessary. The screw cuts its own threads. The use of thread-cutting screws is indicated for sheet metals, softer alloys, etc.

Threading   The cutting of screw threads, either internal or external.

Thread miller   A milling machine designed for thread and worm cutting.

Thread rolling   The formation of screw threads by hardened rolls or dies which roll grooves into a blank and raise enough metal above the surface of the blank to form a thread; such threads are stronger and less expensive than cut threads.

Threads per inch   Refers to thread size. Standard practice fixes the number of threads per inch for any diameter; that is, ½-in diameter, 13 threads per inch; 1-in diameter, 8 threads per inch.; etc.

Thread tool   A lathe tool ground to the shape of the profile of the thread it is to cut.

Throat   The gap in the frame behind the tool in a punching machine, the size of work taken being limited by the depth of the gap.

Through-bolt   A bolt which passes through clearance holes in the pieces to be joined. Clamping actions are secured entirely through the use of a nut.

Thumb out   A wing nut or one so shaped that it can be operated by thumb and forefinger.

Tolerance   Allowable inexactness or error in the dimensions of manufactured machine parts. Also called limit or allowance.

Tool post   A circular post attached to the top of a lathe slide rest for the clamping down of the cutting tools.

Tool, knurling   A tool containing knurls which is held firmly against a piece of revolving work to produce a milled surface, both for ornamentation and to provide a better grip.

Tool post rocker   A finger-like part of steel inserted in the slot of the tool post to give proper adjustment for height of the cutting tool with regard to the work.

Tool steel   Any of the carbon steels or the high-speed steels suitable for use as cutters.

Tooth face   The surface of a machine-tool cutter on which the chip impinges as it is cut from the work.

Trimming dies   Dies used to remove the superfluous metal left around the edges of many kinds of drawn or formed work.

T slot   A recessed, undercut slot made in the table of a milling, planing, or other machine, to receive the head of a T bolt. The slot permits the easy adjustment of the bolt to the position desired.

T-slot cutter   A milling cutter for finishing the wide portion of T slots.

Turret lathe   A lathe with revolving tool head, making possible several operations without removing the tools from the machine.

U clamp   A clamp shaped like the letter u; used for clamping down work on planer beds, etc.

Undercut   Teeth on small gear wheels are said to be undercut when they "thin down" as they approach the root line.

Universal   General; all-reaching; total; entire.

Universal chuck or concentric chuck   A jaw chuck whose jaws are so arranged as to permit simultaneous movement for quick centering of the work.

Universal milling machine   A machine tool having both transverse and longitudinal feed. The work is fed against a revolving cutter. It is similar in appearance to the plain milling machine, the principal difference being that the universal machine has a swivel table.

U.S.F. thread   The United States Form thread has the shape as the National Coarse, U.S.S. thread but differs in pitch.

Vs   Ways shaped like a v, either raised or sunken, to serve as a guide for a movable table or carriage.

Vernier depth gage   A rod-type gage fitted with a vernier and used for checking narrow recessed portions and shoulders or steps of a machine part.

Vertical boring mill   A machine tool with revolving table which carries the work and a slide arrangement which permits both vertical and horizontal feed of the tool. It is especially adapted to a class of work that cannot be easily set up on a lathe.

Vertical lathe   A type of vertical boring mill which carries a side head.

Vise clamps   False jaws for a vise. They are made of brass or copper, and are used over the faces of the hardened steel jaws to prevent bruising the work.

V thread   A screw thread whose section is v-shaped, the included angle being 60 degrees. It is similar to the U.S.S. thread, except that the v thread is sharp at the top and bottom, whereas the others have a flat equal to one-eighth pitch.

Ways   Longitudinal guides upon which the work or a table bearing the work may slide, such as the ways of a lathe.

Wheel dresser   A tool for cleaning, resharpening, and trueing the cutting faces of grinding wheels.

Wheel truing   Any operation on any part of a grinding wheel to balance it or to change its shape to improve its grinding or cutting qualities.

Whitney keys   Square bar keys rounded at both ends.

Whitworth thread   The standard English thread, having rounded tops and bottoms and an included angle of 55 degrees.

Wiggler   A device used on accurate work for exactly locating a center-punch mark on work to be drilled, directly in line with the centerline of the drill spindle. Used also for accurately trueing work in a chuck.

Woodruff key   A semicircular or semielliptical key flattened on the sides, for use in a keyway cut by bringing a rotary cutter against the material.

Worm threads   Threads of the acme type, having an included angle of 29 degress, but usually made deeper than the standard Acme thread.

Index

Abrasives, 322–329
  coated, 323–324
  coatings, 325
  grit sizes, 325
  mounted wheels, 327–329
  storage of, 326
Aligned section drawing, 68
Angle gage blocks, 113–115
Angle of helix, 12
Angular indexing, 301
Angular measurements, 104–108,
  113–115
Annealing, 175
  temperatures for, 129
Arbor, 315
Area calculations, 28–34
Assembly section drawing, 73, 79–
  82, 84
Autocollimator, 113

Belting, high-speed, 418–419
Bench tools, 210–220
  broach, 219, 361–365
  cold chisel, 214
  file (*see* Files)
  hacksaw, 215–216
  hammer, 211
  planer, 283–286
  punch, 355–357
  reamer, 216–218, 259, 333
  scraper, 219–220
  screwdriver, 211–213
  tap (*see* Tap)
  vise, 207, 210
Bench work, 207–220
  layout techniques for, 208–209
  tools for (*see* Bench tools)
Block indexing, 301–306
Blueing, 183–188

Blueprint, 59
Brazing, 133–134
Broaching, 219, 361–365
Broken-out section drawing, 68

Calculators, 2–18
  problem solving with, 5–9
  programmable, 2
  slide rule, 2
  trigonometric functions and, 17–18
Calipers, 91–93
  hermaphrodite, 95–96
  inside, 92–93
  outside, 91–92
Carburizing, 122, 176–177
Case hardening, 122, 176–177
Cast iron, 122
  welding of, 148
Celsius-to-Fahrenheit conversion, 38
Circle solutions, 32–33, 36
Cold chisel, 214
Collet, 317–318
Compound indexing, 300
Computer programming, 18–24
Conversion, temperature, 38
Conversion factors, 39–42
Critical temperature, 174
Cutting fluids, 195–205
  applying, 202
  in drawing, 204
  in grinding, 203–204
  selection of, 196–201
Cutting speed, 5
Cutting tools, 234, 251–259
  fashioning lathe tools, 253–258
  lathe, 251–253
  milling cutter, 258–259

Cutting tools (*Cont.*):
  reamer, 216–218, 259, 333
Cylinder solutions, 34

Damascening, 181–183
Dial indicators, 111–112
Die, 263–268
Differential indexing, 300–301
Dimensioning, 75
Direct indexing, 300
Dividing head (*see* Indexing head)
Drawings, 59–90
  aligned section, 68
  assembly section, 73, 79–82, 84
  blueprint, 59
  broken-out section, 68
  diagram, 63
  dimensioning on, 75
  full section, 68
  half section, 68
  jig and fixture, 75–79
  line conventions for, 61, 67
  orthographic projections, 62–63
  pictorial, 60
  removed section, 68
  revolved section, 68
  schedule representation on, 73
  sectional view, 64–73
  symbols on, 73–74
  working, 82–90
Drills, 271–282
  lathe-mounted, 281–282
  radial, 274–281
  twist, 271–274
Drives, 412–414

Electric motors, 408–412
Electroplating, 190–191

Ellipse solutions, 34
Energy, 407
Engine turning, 181–183
Equations, 25–27

Fahrenheit-to-Celsius conversion, 38
Files, 393–399
  care of, 404–405
  curved-tooth, 397–398
  cut of, 394–396
  length of, 393
  machinist's, 396
  mill, 396
  selection of, 398–399
  styles of, 394
  Swiss pattern, 397
  (*See also* Filing)
Filing, 399–405
  draw, 401–402
  lathe, 237, 402–404
  straight, 399–401
  (*See also* Files)
Forging, 175
Full section drawing, 68

Gage blocks, 119
Gears, 373–377
  teeth on, 13
Grease, 205
Grinding, 321–340
  abrasives for (*see* Abrasives)
  cutting fluids in, 203–204
  external, 333–334
  hardened steel part, 332–333
  internal, 334–335
  lathe, 331–333
  precision, 329–331

Grinding (*Cont.*):
  reamers, 333
  surface, 336–340
  valve, 335

Hacksaw, 215–216
Half section drawing, 68
Hammer, 211
Hermaphrodite caliper, 95–96

Indexing, 299–312
  angular, 301
  block, 301–306
  compound, 300
  differential, 300–301
  direct, 300
  (*See also* Indexing head)
Indexing head, 299–312
  adjusting, 311–312
  mounting, 308–311
  sector arms on, 306–308
  (*See also* Indexing)
Inside caliper, 92–93
International System of Units (SI),
    43–58
  of force, 56
  of mass, 55–56
  prefix rules for, 48–49
  prefix symbols, 48–49
  pronunciation rules for, 48, 54–
    55
  typewriting recommendations for,
    56–57
  of weight, 55
  writing of metric quantities, 44–
    54
Iron, 121
  cast, 122, 148

Jeweling, 181–183

Knurling, 236

Lapping, 341–353
 compounds for, 341–346
 hand, 347
 lathe, 237
 machine, 347–348
 of rifle barrels, 348–353
Lathes, 221–250
 adapters for, 318–319
 center alignment of, 227–232
 center rest in, 238
 drilling with, 281–282
 filing with, 237, 402–404
 follower rest in, 238
 grinding with, 331–333
 installation of, 224–227
 knurling with, 236
 lapping with, 237
 milling with, 240, 296–298
 parts of, 222–224
 polishing with, 237
 spring winding with, 237–238
 tail-stock setover in, 7–9, 20,
  242–248
 taper turning in (*see* Taper
  turning)
 thread cutting and, 390–392
 workpiece placement in, 232
 (*See also* Cutting tools)
Lubricants, 205–206

Mandrel, 313–315
Measurements, 91–119
 angle gage blocks, 113–115
 angular, 104–108, 113–115
 caliper (*see* Calipers)

Measurements (*Cont.*):
 dial indicators, 111–112
 gage blocks, 119
 micrometer (*see* Micrometers)
 thread, 115–119
 vernier height gage, 108–111
 (*See also* International System of
  Units)
Metal spinning, 367–371
 tools for, 368–371
Metric system of units (*see*
  International System of Units)
Micrometers, 97–104
 blade, 102
 depth, 104
 hub, 103
 inside, 104
 metric, 98
 point, 101
 screw thread, 101–102
 tubing-type, 101
 vernier, 99–100
Milling, 289–298
 keyway cutting, 292–296
 use of lathe for, 240, 296–298
 (*See also* Indexing; Indexing
  head)
Motor selection, 415–423
 duty cycle, 420–421
 high-speed belting, 418–419
 inertial load and, 422–423
 load factor, 416
 mounting designs, 417–418
 power supply, 416
 torque determination, 415–416
 (*See also* Drives)

Nickel plating, 189
 (*See also* Blueing)
Nitriding, 176

Oil, 205–206
Orthographic projection, 62–63
Outside caliper, 91–92

Parallelepiped solutions, 35
Parallelogram solutions, 30
Parkerizing, 188–189
Pictorial drawing, 60
Planer, 283–286
Plating, 189–191
    electroplating, 190–191
    nickel, 189
    (See also Blueing)
Point of recalescence, 174
Polishing, 192–194
    lathe, 237
Polygon solutions, 31–32
Power, 407–409
    electrical, 408–409
    mechanical, 407–408
Prism solutions, 35
Projection comparator, 116
Protractor, 107–108
Punch, 355–357

Radial drill, 274–281
Reamer, 216–218, 259, 333
Rectangle solutions, 30
Removed section drawing, 68
Revolved section drawing, 68
Rifle barrel lapping, 348–353
Rockwell hardness, 126–127, 178–
    179

Scleroscope hardness tester, 179–
    180

Scraper, 219–220
Screw pitch gage, 116
Screwdriver, 211–213
Shaper, 283–287
Shears, 357–359
SI (see International System of
    Units)
Slide rule calculator, 2
Snips, 357–359
Soldering, 131–134
    fluxes for, 132
    hard, 133–134
    soft, 132–133
Sphere solutions, 35
Stainless steel, 122, 123, 125–126
Steel, 121–129
    alloy, 122, 125
    carbon, 124
    carbon content of, 122, 124
    heat treatment of (see Steel heat
        treatment)
    high-carbon, 124
    high-strength, 125
    low-carbon, 122
    spark testing of, 123
    stainless, 122, 123, 125–126
    tool-and-die, 126–129
Steel heat treatment, 173–180
    annealing, 175
    carburizing, 122, 176–177
    flame-hardening, 176
    forging, 175
    furnaces for, 177
    hardening, 173–174
    hardness testing for, 178–180
    nitriding, 176
    normalizing, 175
    tempering, 126–129, 174
    thermocouple pyrometer and,
        177–178
Steel rule, 93–95

Tail-stock setover, 7–9, 20, 242–248
Tap, 215, 261–263
  extractor for, 269
Taper angle, 9–16
Taper turning, 240–250
  tail-stock setover for, 7–9, 20, 242–248
  taper attachment method, 248–250
Temperature, critical, 174
Temperature conversion, 38
Tempering, 126–129, 174
Thermocouple pyrometer, 177–178
Threading, 379–392
  compound rest and, 390
  lathe cutting and, 390–392
  thread dial and, 389–390
  (*See also* Threads)
Threads:
  Acme, 382
  measurement of, 115–119
  metric, 269, 386–387
  multiple, 387–389
  square, 384–386
  v, 381
  Whitworth, 386
  (*See also* Threading)
Torque, 415–416
Trapezium solutions, 31
Trapezoid solutions, 30

Triangle solutions, 14–15, 28–30
Twist drill, 271–274

Valve grinding, 335
Vernier height gage, 108–111
Vernier micrometer, 99–100
Vise, 207, 210
Volume solutions, 34–36

Wedge solutions, 36
Welding, 134–171
  arc characteristics, 134–136
  arc welder cutting, 149
  butt, 137–139
  cast iron, 148
  electrode selection for, 149
  fillet, 139–140
  hard surfacing, 144–148
  metals for, 136
  overhead, 143–144
  safety in, 149, 169–171
  sheet metal, 144
  vertical, 141–143
  vertical-down, 142–143
  vertical-up, 141–142
Work, 407
Working drawing, 82–90

About the Author

John Traister has had many years of experience as a machine shop foreman. He has worked in a variety of shops, doing metalworking tasks on the full range of machines from lathes to milling. He is also the widely known author of books on a variety of industrial topics, including *Electrical Design for Building Construction,* 2d. ed., and *Design and Application of Security/Fire Alarm Systems,* both published by McGraw-Hill.